Survival Skills
for **Scientists**

Federico Rosei • Tudor Johnston

Institut National de la Recherche Scientifique
Université du Québec, Montréal, Canada

Survival Skills
for Scientists

Imperial College Press

Published by

Imperial College Press
57 Shelton Street
Covent Garden
London WC2H 9HE

Distributed by

World Scientific Publishing Co. Pte. Ltd.
5 Toh Tuck Link, Singapore 596224
USA office: 27 Warren Street, Suite 401-402, Hackensack, NJ 07601
UK office: 57 Shelton Street, Covent Garden, London WC2H 9HE

British Library Cataloguing-in-Publication Data
A catalogue record for this book is available from the British Library.

First published 2006
Reprinted 2007

SURVIVAL SKILLS FOR SCIENTISTS

ISBN-13 978-1-86094-640-0
ISBN-10 1-86094-640-2
ISBN-13 978-1-86094-641-7 (pbk)
ISBN-10 1-86094-641-0 (pbk)

Printed in Singapore.

A Sabine, con tutto il mio amore

To Anne, for everything

Preface

This book is a direct reflection of the experiences of the authors who are scientists, not professional writers and so, to help the reader place the book in context, we introduce ourselves.

Federico Rosei was born in Rome (Italy) on March 27th, 1972. He received a "Laurea" degree in Physics in February 1996, and a Ph.D. in Physics in February 2001, both from the University of Rome "La Sapienza". From October 1996 to December 1997 he was an Officer in the Italian Navy. He continued his scientific career as a Post-doctoral Fellow and Marie Curie Fellow at the University of Aarhus (DK), from November 2000 to April 2002. In May 2002 he joined the faculty at Institut National de la Recherche Scientifique (INRS) Enérgie, Matériaux et Télécommunications, University of Quebec in Varennes, as Assistant Professor. After two years, he has been promoted to Associate Professor with tenure (since June 2004). Since October 2003, Dr. Rosei has held the Canada Research Chair in Nanostructured Organic and Inorganic Materials. He has co–authored over 50 papers and has given over 60 Invited, Keynote and Plenary Talks at international conferences. His research is on nanostructured materials, and on controlling their size, shape, composition, stability and positioning, when grown on suitable substrates. He is also known for his studies on the assembly and dynamics of molecular aggregates on surfaces.

Tudor Johnston was born in Montreal (Canada) on January 17th, 1932. Graduating from McGill University in 1953 in Engineering Physics, and obtaining a Ph.D. (in Engineering) at Cambridge University in 1958, he transformed himself into a plasma theorist at the RCA Research Laboratories in Montreal, where he remained for ten years. After a short stint in Texas at the University of Houston, he returned to Montreal to INRS Enérgie, Matériaux et Télécommunications at Varennes. He has published over 150 papers and co-authored one monograph in plasma physics, on many plasmas (ranging from space plasmas with 30 kilometer wavelengths to solid-density plasmas with wavelengths of 10 nanometers). In this way he has become familiar with industrial research, the Canadian scientific establishment structures and operations, both English and North American academia, and acquired some informal acquaintance with several science establishments in France and in England.

As one can immediately infer, this book is the result of a collaboration between two people, one who is beginning to open into full flower and the other who has most of the best years behind him. In this partnership we intend to combine the best of the points of view of the young professor (Rosei) who has recently experienced many of the early rites of passage of science (and so remembers them clearly) with the more reflective and experienced viewpoint of the other (Johnston) who has watched and played a part in many such events over more than forty years.

The usual "voice" of this book is the combined consensus voice of the two. Sometimes, however, we have agreed to disagree, putting both points of view and leaving the choice to the reader, in which case the reader is explicitly told (identified there as **Federico** or as **Tudor**). Sometimes one of us (usually, but not always, Federico) launches into a personal comment or an anecdote. (Most of Tudor's anecdotes would be somewhat dated.) Finally, with respect to voices, unless specifically stated to be otherwise, the illustrative personal anecdotes relating to events fairly early in a scientist's career are all contributed by Federico Rosei, and so the pronoun "I" will make an occasional appearance there.

To sum up, when the voice is not of us both together, that voice will be explicitly identified; when no voice is explicitly identified it is our common voice. (In computer jargon, the common voice is the "default option".)

We have both looked at everything that is set down here, so this whole book does legitimately bear the overall imprimatur of both intense youth and more reflective experience.

Disclaimer

The advice and opinions contained here naturally reflect solely those of the authors and we take full responsibility for them. Being our personal ideas and recipes for success, we naturally do not claim that they are universal. They are certainly not to be intended as scientific laws or as infallible pronouncements *ex cathedra*.

We do hope that, fallible though it may be, the advice to be found here will prove useful to the readers, if only to provoke them into formulating their own views and responses. In either case our principal goal will have been attained. That goal is to help our readers to navigate the unavoidable difficulties that most young scientists encounter, to evade the more obvious pitfalls and to avoid most common mistakes in situations that all of us will encounter or have encountered in one form or another in our careers in science.

The illustrative anecdotes are not invented and do treat actual events. For this reason it was decided to use for each just a one-letter "name" (rather than using actual initials), to better protect the identity of the people involved. The point of view in the anecdote is that of the narrator, who thus accepts the responsibilities for any misinterpretation there may have been.

Acknowledgements

The idea of writing this book was conceived during the preparation of a graduate course on Survival Skills for Science, offered at INRS-EMT in the fall term of 2003, by Federico Rosei together with our common colleague, Alain Pignolet. We are extremely grateful to Alain for his key contributions in that course. Elements of this work were presented at various times in various places and we are grateful to those who proffered advice, comments and encouragement.

Professor Peter McBreen was kind enough to read this book in manuscript form and offer some invaluable suggestions.

In addition to these acknowledgements Federico Rosei has his own individual acknowledgements.

Federico: — I am grateful to Peter J. Feibelman for the great advice I found in his book, since I am confident that it helped me significantly in the early stages of my career. I hope the present book will serve as a complement to his excellent perspective. I thank colleagues Roberto Morandotti and Mohamed Chaker for very helpful discussions, and also Travis Metcalfe for introducing me to Feibelman's book, and for many entertaining and stimulating discussions on professional development and on how to survive in science. I acknowledge Sabine and Chantal Minsky as well as my parents, Silvana and Renzo Rosei, for reading this manuscript thoroughly and critically, and for many suggestions on how to improve it.

We welcome your comments!

In case there is going to be a further edition, we would greatly benefit from the reader's (constructive) comments and criticism. Thus we encourage you to send us your thoughts (as well as lists of typos), using various options, such as:

Email:
rosei@emt.inrs.ca
johnston@emt.inrs.ca

Ordinary mail:
Professor Federico Rosei and Professor Tudor Johnston
INRS-EMT, Univ. du Quebec
1650 Boul. Lionel Boulet
J3X 1S2 Varennes (QC) Canada J3X 1S2

Introduction

This book is intended as a tactical guide for scientists who will be faced with many career choices in the next few years. We do not presume here to give you advice about how to do the science itself. We remind you of the things to keep in mind when you are faced with various choices. Among these career choices are obvious ones of post-graduate school, thesis advisor, where to go as a post-doctoral fellow and with what kind of director, applying for faculty or research positions, but also where to publish, how to conduct yourself in the eternal three-player game (between author, referee and editor) for publication, and much more. While chiefly aimed at the large community of research scientists whose success depends directly on the research that they publish in peer-reviewed literature, there is much that is useful for those looking to work in laboratories where satisfying your supervisor is more important than satisfying some anonymous referees. Although we address directly the young scientists on their way up the ladder, there is also much here that the more senior scientists can gain from this book, particularly in helping to understand the professional preoccupations of younger colleagues and employees.

The genesis of this book (and of the graduate course from which it sprang) was our realization that most of what we have learned "on the job" in terms of career skills is simply not available in published form except partially in some scattered fragments. (Far more advice can be found on buying a car or a house or starting a business!) Our aim is to help the young scientist in making the many choices that fall under the headings of the career strategy, tactics and planning considerations which can be vital carving your career path in science. We wish to contribute to filling this gap.

As an exception to this last statement, there is already an excellent book by Peter J. Feibelman[a] on advice to the young scientist. However, his book focuses mostly on the situation in the U.S. We have extended this way of looking at things both geographically and also far beyond the post-doctoral phase. Moreover, although Feibelman does compare academia, industry and government laboratories, he appears to have a clear preference for and has more space devoted to the latter (perhaps because he works in a government lab?). On the other hand, we aim more at the academic milieu, which is our personal preference. We express here our gratitude for the insights and approach afforded by Feibelman and for the thematic inspiration his work gave us. The reader of this book will find Feibelman's book very useful for the aspects that it does treat, and for an interesting contrast to our own remarks on the topics we both discuss.

If scientists had career agent/managers our book summarizes the kind of advice they would give. *This book is thus aimed at helping you to become the agent/manager of your own career.*

Without these managerial skills and their application, your progress in the world of science will be left far more to chance than it should be. You will do better as a scientist who makes things happen, rather than as a scientist to whom things happen.

Most scientists (and nearly all those beginning in science) attend to career matters above only when faced with a deadline of some sort ("My thesis is finished, now what?", "My fellowship will be up soon, what next?" "Oops! That deadline is the end of this week!"). This means that only a minimum of planning can be done. You can do far better if you take the trouble to tend to your career at least on a weekly basis. The first and basic piece of advice, the *"zeroth law"* of scientific survival is *pay attention to your scientific career.*

We have articulated three basic themes which can be likened to the three legs of a good tripod in that all the legs are indispensable. In the natural order of application: the first law is *"Know thyself."* (This ancient

---

[a] P.J. Feibelman, *A Ph.D. Is Not Enough: A Guide to Survival in Science* (you can find it on Amazon.com). If you are serious about pursuing a scientific career, Feibelman's book provides an excellent complement to this book.

Greek aphorism (Greek: Γνῶθι Σεαυτόν or gnothi seauton) was inscribed in golden letters at the lintel of the entrance to the Temple of Apollo at Delphi.) To this we add two more things to know. The second law is "*Know your tradecraft.*" and the third law is "*Know thy neighbor.*" What do we actually mean by these general maxims?

"*Know thyself.*" Before you apply the other "laws" you should know the client — yourself — not only as you might know yourself in life, but as a player in the "game of science." Pay attention to yourself, to your strengths and weaknesses and how to improve them. Do not try to be the perfect scientist, try to be the best scientist you can be. (This book is for you, not those others.)

"*Know thy tradecraft.*" In many spy novels (and in particular in those of John Le Carré) the useful word "tradecraft" means the technique of organizing the mail-drops, the packaging and sending of the information and so on. The "published science research game" includes not only the actual science you do but much more. This "much more" is what we are here terming "tradecraft". At the basic level, "tradecraft" means the craft of writing papers that people want to read, constructing seminars that are fun to hear and to give, learning how to perform in an interview and much, much more. At this level you are targeting (so to speak) other scientists at large rather than any scientists in particular. At the more advanced levels to follow, "tradecraft" is developed and deepened to include the art of initiating individual contacts with others to advance your professional ambition and your science, which requires tailoring your tactics to the occasion and to someone or to several people.

"*Know thy neighbor.*" To succeed on managing your interactions with others, you must pay attention to the people with whom you will be interacting. Work diligently at putting yourself in their shoes so that you can do a better job of tailoring your impact on them. These will include co-workers, supervisors, the listeners to your seminars and oral presentations of many kinds, the readers of your publications, grant applications and job applications, and many more. When you come to be responsible for directing others these skills will be invaluable in helping you to manage the research of others.

With these three "laws" in mind you are ready for another fundamental piece of advice for a successful scientific career. This basic concept is simple enough to say, but it takes a disciplined, sustained effort to put it into practice. It simply is: *plan ahead to do the best you can in your scientific career*. The theme of this book is the working out of this basic piece of advice in various aspects of your scientific activity. Like much good advice it seems trivial, but, as is so often the case, it is rarely carried through in practice. *Do not plan just your scientific work; you should also plan your career in science.*

Think of the science to which you are becoming addicted as an actively turbulent river down which you are being swept. There are dangerous rapids which you can use to make better progress at some risk, safe but stagnant pools where no progress is being made, disastrous falls and the like. Clearly you do better if, instead of just letting yourself be carried by the current, you plan ahead as much as is feasible and also set yourself up to profit by unforeseen opportunities and also to take action before being overwhelmed by disaster. The same concepts apply to your career in science. Those who plan ahead, and who are ready to profit by opportunity, are far more likely to be able to do the science that they would most enjoy, and to have more control over how it is done, than those who do not. Those who let things happen to them will wind up becoming servants of those who make things happen.

So think carefully about your short-term plans, as well as your medium-term and long-term plans. Think as far ahead as ten years in the future. Do not worry about whether your plans or dreams will come true. Most young people are content with taking their lives one day at a time, without any attempt at long-term planning. This is also true for young scientists, who are often ill at ease (almost guilty) about planning their careers (or even about planning what they will do next year). Unfortunately, they do not realize the extent of the threat this carelessness poses to their future. Scientific research is a world of opportunities, and, of course, of competition to take advantage of them. In this sense, careful planning can be extremely helpful, especially in terms of having an alert and prepared mind.

This does not mean that, in order to do the science you would like to do, you must try to become a local despot of science like a military officer or the coach of an American football team (although some few may find their success that way). Between the solitary scientist (with perhaps a graduate student or two, and perhaps a post-doc), and the "czar of all the Russias" commanding a vast army of researchers, there are many successful modes of operation and collaboration which are more democratic, informal and more fun than either of these extremes. The point is to be able to find out what mode of science suits you and your science best, and how to be successful in creating a viable niche for that mode, and thus attain a happy equilibrium between your life and your science.

It is true that experience — your own and that of the people around you — will be your best teacher, provided that the experience itself is not fatal to your career. (This is a deliberate echo of Nietzsche "That which does not kill us makes us strong.") The risk and trouble is, however, much reduced if you learn as much as you can from other people's experiences first, rather than from your own lessons painfully learned from disasters. And that is what we are offering here. Learn as much as you can from us to reduce the pain to yourself.

Of course this guide is not perfection, merely our opinion, so it should be read critically, just as you would read a scientific article. We do not offer all the solutions, but display options try to get you to think seriously and objectively about your future. Thinking in this way about the problems that lie ahead is the first and necessary step to figure out ways to consider these problems in advance and thus to have your plans ready before the problems arrive.

What we write essentially stems from our direct experience or from the experience of colleagues and friends. We try to draw examples from real life whenever feasible; you may thus more easily identify with situations you have lived or at least heard about. In doing so, we do not mean to be discouraging or enlightening, but simply to provide some information and advice that may be useful in your career, and perhaps even in your life in general. You may find this career strategy somewhat cynical, and perhaps it is, however we believe that is far better than relying on hope and the kindness of others.

To repeat, the central idea or message of this book is to promote awareness of what to expect in pursuing a scientific career, to stimulate you to ask yourself many career questions, and to try to make plans in advance. If you do all this, we will consider it a great success, even if your way differs a lot from ours. If you do only a modest fraction of what is presented here, we believe you will profit much. (In fact we feel the utility of the book is such that someone who is so well established as to not feel the need for personal use will likely find it worth recommending for the graduate students or postdoctoral fellows.) Even if all you do is to read "Survival Skills for Scientists" through just once thoughtfully, we think that you find the work moderately enjoyable, while your friends and colleagues who work more on the points we raise may find it really useful.

The book is organized as follows. We start with the "Basics" of pursuing a scientific career. In particular, we focus first (Chapter 1) on the fundamental questions: what is your goal in becoming a scientist? or even more basically, why do you want to do science? and in what general style do you want to do your science (lone wolf, team player etc., academia, government laboratory, industrial research)? and how may cultural and geographic differences affect your choices and career? Other topics are finding a mentor, language skills, patience, fighting against the odds, equal opportunities, diversification, working in Asia etc.

Having examined yourself and decided on what style of research you want to pursue and the overall work choices, it is now appropriate to discuss in general terms in Chapter 2 how to begin to make choices of where you want to work, and how to succeed with these choices and how to keep yourself open for more. This will include how to gather the information to help you learn what opportunities arise and their relevant deadlines.

In Chapter 3 we discuss what might be termed the actual "game of science" itself: its ecology, how peer review works in practice, how things can go wrong, ethical issues etc. These last can involve getting proper credit for your work, intellectual property rights and patents, and the like.

Now that the "rules of the game" have been established, the next topic (Chapter 4) is naturally how to present and "package" your work so as to become as well-known as you and your work deserve. The topics here include publishing tactics (where and how, journal citation indices and impact factors come into play) and conferences (which and how, oral presentations or posters).

Finally in Chapter 5 we take up the topic of the scientific writing itself, the basic concepts for a peer-reviewed publication, a thesis, the scientific heart of a research proposal. It is also important to know how to write your CV in the light of the context in which it will be used, writing applications for scholarships, fellowships and of course research proposals. Here also are the different types of more ephemeral presentations, the oral presentations: conferences, seminars, job interviews, and conference posters. Getting known in your ideas includes learning how to present your science in grant applications and how that differs (less than you would imagine) from a peer-reviewed paper.

Chapter 6 contains some cautionary tales, while some overall conclusions and an envoi form Chapter 7.

Contents

Chapter 1

Basic Choices

Sections of this Chapter

1.1   "Know thyself" so you can set realizable goals

Make the effort to "Know thyself"[a] so that your goals are realistic and will indeed satisfy you when you attain them. What should be your career goals as a scientist? This should include not just what you would like to achieve as a scientist, but how to advance in your career so as to have the means to do what you want. Asking yourself this question openly, critically and realistically at each stage of your career (preferably well before the next stage is to begin) is extremely important. It may save you from a lot of trouble and frustration, later on. Of course you should not forget to ask yourself this basic question from time to time later in your development as a scientist (say every few months at least), and not

---

[a]"Know Thyself." This famous Greek maxim is attributed to any number of ancient Greek philosophers, including the great Socrates. However, according to the ancient historian Plutarch, "Know Thyself" was originally the admonition Γνῶθι σεαυτόν pronounced "*Gnothi se auton*" ("Know Thyself") inscribed on the Sun god Apollo's Oracle of Delphi temple in ancient Greece. Plutarch should know about the inscription on the Oracle, since he was once one of its caretakers. In deference to Socrates, it is known that Apollo's Oracle of Delphi identified him as being the wisest of all men.

just at the moment when you begin to think of a career change. Update your self-examination and your goals and not just your *resumé*. Your position is not unlike a young artist in the Renaissance. It was not enough to develop your skills and your vision, it was equally necessary to plan to get into the good graces of a patron to support you in the pursuit of your art.

Did you ever ask yourself, what are the goals of a scientist in general, and what should be your own goals? Remember that the whole point is to help you become what you want to be.

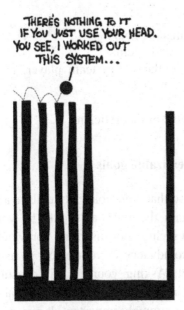

Success is of course not guaranteed, as is illustrated in a brilliant 1966 cartoon by Jack Wohl,[b] shown here. (He had earlier made the deep discovery that one can often get away by representing people with simple shapes like circles.) However do not be deterred by this image from planning at all. After all, if you form a good plan and act on it, you will usually do better than if you do not plan at all. All the message that you should draw from this is that no plan is likely to work for ever.

As you will see, we found that much of what we have to say has its humorous side with some sources[1] coming up quite frequently: the cartoonist Sidney Harris, Robert Weber who produced two impressive anthologies of science humor and Stanley Krantz who has done something similar for mathematics.

Even more basic questions are *"Why* do you want to become a scientist? *What* will you be able to contribute?" If you have not yet asked yourself these simple questions, you should definitely do that immediately. It does not make much sense to try to become a scientist if you do not know what your deep objectives are.

---

[b]Jack Wohl, *The Conformers*, P-S Books, Pocket Books New York, NY (1966).

Unfortunately many young people keep on with a topic because it feels good to do, but when it becomes difficult (as it will), they can be quite shocked, almost like falling out of love. Thinking of the early love affair with science that we have all had, this really means thinking about whether you want to marry your lover and stay together for life. If you do not come up with what you think is a reasonable answer, it is wiser *not* to pursue a scientific career before too much time is wasted. (However, you should by all means at least finish your degree if you are still a student, and then look for a different way to use your talents.)

Many (including ourselves) believe that the main goal of a pure scientist is to investigate the laws of nature, and to provide new knowledge and insight into the physical, chemical and biological processes of systems which either exist already or which you create.

"Real scientists" are those who burn with the desire to unravel the mysteries of nature. The other aspects of the job are essentially the means to achieve this end. A real scientist, if offered the same salary by an intelligent philanthropist, but without the necessity of having to do anything specific for this money, would choose to carry on in the same way as at present. A real scientist is, in fact, addicted to science.

While personal remuneration is not the primary motivation to the real scientist, money is not irrelevant, though in an unusual sense. To the real scientist "money" means the funds to pursue the desired research. Of course, without enough funding it is very hard and challenging (though not altogether impossible) to do good science, so the funding is a necessary (but never completely adequate) means for an end. (This said, it is true that some people in research do consider money, in the sense of funding, an end in itself.[c])

Having a lot of funding is not enough, however. There are many scientists (and most of us can easily bring a few to mind), who have substantial money flowing into their research accounts, and yet produce disappointingly little in terms of interesting science of high quality. Their groups are often so large that they have a hard time managing them, and

---

[c]This seems to be especially true in North America. In Europe, your peer's respect is earned by means of a long publication list in prestigious scientific journals. It is rare (though not unheard of) for scientists in Europe to boast about their level of funding.

their productivity (in the sense of published papers per year and per member of the group) suffers enormously from this. The point is that it is the new, original ideas that are the most important ingredients for successful science. There are also colleagues who have relatively little money, and yet their output — perhaps normalized with respect to the amount of research money they have — is incredibly high. (This is not limited to theoreticians, particularly analytical ones, who only need good ideas, a pen and some paper, besides some travel money to go to conferences. Some experimentalists, for example in Italy, Federico's home country, do wonders with the little funding they have.)

If your basic real goal is to become rich[d] and famous, not in terms of your research accounts but in terms of your own bank account, we strongly suggest that you pick another career that will let you attain those goals and fulfill your aspirations in a simpler and faster way. (Perhaps you should try to become a techno-industrial wizard like the co-founders of Google.) Since there are already more than enough *prima donnas* and empire-builders in science who are really driving for the best science; there is no need for more who are mistakenly counting on science to become rich. True, the odd scientist does win the Nobel Prize, and may thus become famous[e] and moderately rich. A very small proportion of scientists who founded companies manage to do rather well and earn the scientist considerable sums. However these people are nearly always

---

[d]One of Sidney Harris' cartoons (as shown on p. 90 of his *Chalk Up Another One*, Rutgers University Press (1992)) shows a dispirited scientist at a cocktail party saying, "My big mistake was going into cosmology just for the money."

[e]The concept of fame here is very relative. If you meet someone on the street, and ask them who, say, Tiger Woods is, they will probably know a lot about him, including some of his greatest feats, some details of his personal life and advertising. On the other hand, if you ask who Albert Einstein was, they may have a vague notion of a funny scientist who said that *everything is relative*. They will probably also joke about it, saying that it is quite obvious to them that everything is relative, and that giving a Nobel Prize to someone for such a trivial discovery is an exaggeration. (In case you did not know, Einstein was awarded one Nobel Prize in Physics, in 1921, for his description of the photoelectric effect (published in 1905). By contrast, both his theories of relativity, the special and the general one, were not widely accepted by the senior scientific community until long after he formulated them. Now General Relativity is an essential element in the programme used for Global Position Satellite (GPS) operation, so we all use Einstein's general relativity, whether we know it or not!)

"real scientists" (in the sense that there may be exceptions of which we have no knowledge) who encountered an opportunity and managed to profit by it. Going into science in the hopes of finding such an opportunity is a poor bet and not worth considering as a realistic possibility. (See below, hand drawn by Sabine A. Minsky.)

Even if you are unlikely to become rich in science, and if you are likely to work long hours, at least a scientific career will not be a work of boring and safe routine, like an office job. In an active research program with several components it is extremely rare to have two days that look alike. If you are upset without a steady routine, do not try to be a successful scientist. The activities that a scientist is engaged in are so various and different that they are rarely repetitive (although some of them are boring, of course). You should expect to work long hours, perhaps even spend many weekends working (especially if you work in academia), and hopefully you should even take pleasure in it. Generally speaking, this is definitely *not* the type of job in which you work from nine to five each working day and then go home, forget about all your problems at work and concentrate on your family and/or your hobbies. (A completely free weekend may be a rarity.)

To repeat, if you are not prepared for commitment far beyond nine-to-five routine, you should seriously think of looking into different job prospects.

In spite of what has just been said, there are exceptions, but ones for which this book is irrelevant, since it is intended as an aid for those who are determined on climbing the ladder of success in science. These exceptions are people who are settled in the science niches in which a routine "nine-to-five" scientist without some considerable "fire in the belly" can operate in a limited way. If we have "alpha" scientists and "beta" scientists perhaps these are "delta" scientists who are not even thinking of being "gamma" (= sub-"beta") scientists. The point is that these positions are just that, niches, and not way stations on a path to success in science.

In taking such a position in a way that is more or less permanent, you should realize that you would be essentially joining the ranks of the many middle and lower level scientists (often called research associates) in the world who are essentially well-educated technicians. Such people have reached a stable niche position. Their names are on the papers (naturally in subordinate positions); they may lead small technical projects, and they have accepted that they will work on projects decided by others. By way of compensation and to use energy which did not seem to work in pure science, they often become happily addicted to some pursuit outside science.

This is a perfectly valid role for those that choose it, but it should be a conscious choice, "I really like science, but I like my family and my other activities just as much or more, and so I willingly choose this useful, enjoyable but secondary role as my compromise." What you should not do is to try vainly to be a top (or at least excellent) scientist (an alpha or beta scientist so to speak) and then settle resentfully into such a tertiary position, working in a "nine-to-five" work-to-rule as an implicit protest at the loss of an early dream. In this context "Know thyself" means to accept that your fate or your true desire is perhaps other than what you had in mind as an undergraduate.

If that is your choice, if you have dropped out of the race and into a non-competitive niche, then this book is only of sociological interest for you, perhaps as an aid to understanding what the "upwardly mobile" people around you are doing.

In essence, then, this book is really addressed to those who want to be highly successful in science. Whether you are successful or not may then depend on external factors such as pure luck, like in most human endeavors (look at Jack Wohl's cartoon on page 2 again), but if you are on this path you should be trying as hard as you can.[f]

If you are a true scientist, your enthusiasm will clearly be displayed when you give a talk, or when you write a scientific article or even a

---

[f]Would it make sense to want, say, to become a professional tennis player, without aspiring to be number one?

grant proposal. Your peers will look up to you with respect, sometimes even with awe, and consider you as a source of inspiration (except that unfortunately large fraction who are jealous or envious). You will get very excited[g] (a science "rush" in fact) when you or your students acquire new results and now understand something that nobody has ever understood before. This excitement and enthusiasm are the true rewards of a scientist, and they make up, or at least should make up, for most of the drawbacks, pitfalls and "sacrifices" that come with the job. It is important to keep this in mind to face the rough times that undoubtedly lie ahead. A scientific career is tough, and full of unknowns, especially at the beginning. However, very few people love their jobs as much as scientists do, and this is to be considered as an enormous fringe benefit.

A scientific career can be extremely demanding in terms of the total number of hours you work and the little vacation you feel comfortable taking. There is always something interesting to do, and new ideas and challenges turn up at an astonishing rate. At the same time, you will be traveling frequently to attend meetings and conferences, and give seminars on your work, which may seem pleasant for you, but which will not make your family terribly happy (if you managed to find the time to start a family in the first place, that is), unless you systematically bring your family with you, of course (which is rarely practical, unfortunately).

In fact, if you consider these demands on your time as sacrifices, and you are not willing to accept them to advance your career, we emphasize yet again that you should seriously consider looking for another type of employment. Most good scientists tend to work 10-12 hours per day, often including weekends.[h] They feel comfortable with this situation, because they literally love their job and have loads of fun while working.

---

[g]This is the same excitement that seizes children when they receive a new toy. We have seen this happen to colleagues a great number of times!

[h]It sometimes feels somewhat frustrating to know that if scientists were being paid by the hour instead of monthly, they would soon become rich. However one of the reasons for choosing this career in the first place is that money is *not* at the top of our priority list. Indeed, if money is important for you, you are certainly better off choosing a different career. There are other important fringe benefits in a scientific career, and perhaps the most important one is that you actually have fun while you are working. In fact, it is not uncommon for me to describe my job as "Playing with atoms and molecules" (not to my peers, who know it only too well, but to the layman). Just imagine that I get paid for it!

Being a scientist often means that you learn new things almost on a daily basis, and this can be very interesting, stimulating and challenging.[i] They are very few people who are efficient and intense enough to work only 8-9 hours per day, five days a week, and yet display sufficient scientific productivity to keep them in good standing in the community. These remarks are not meant to discourage these efficient scientists, but just to point out that they tend to represent a minority in our community, so you are *a priori* unlikely to be one of their number (Are you really so efficient? Know thyself!).

At the same time, if you choose to work in academia, once you reach the level of a professorship, you will have to become a *manager* besides being a scientist. You will need to attract funds to carry out your work, and you will have to manage them properly. If you are publicly funded, you will be asked to report fairly regularly to the funding agency on how you spent the taxpayer's money. In essence, a scientific job is often multifaceted, and will expose you to many different realms of human endeavor.

Once again, all this is really a matter of many personal choices. As remarked at the outset, *"Know yourself!"* To repeat our refrain, if you know yourself well, this knowledge will help you in making the right choices, and in the long run you will be a happier person. This does not apply only to your scientific career, but to your choices at the restaurant.[j] (It would take a braver pair of authors than ourselves to suggest that this be applied to a choice of a life partner, sensible thought it might be, so we leave this concept at the restaurant from whence it came.)

A lot of young people tend to be undecided and drift along hoping that something good will turn up, like a swimmer in a large river hoping a useful boat will float to within their grasp. They avoid planning, partly because they are frightened of the future (and perhaps because they are

---

[i]**Federico:** if you think that I work too much, you just have to look at what other *successful* colleagues are doing, to realize that I am not exaggerating (although I admit that finding an objective measure of success is not at all obvious). Once in a while I do take the odd weekend off, or even the odd vacation. However, this is the working style that I learned from my father, who is also a physics professor: working long continuous hours helps you keep a high level of concentration. Unless you are as bright as Albert Einstein, this seems to be one of the best approaches to succeed in this job.

[j]If you don't like chicken, don't order it!

secretly hoping that they will suddenly stumble on a gold mine). They do not yet know themselves well enough to be able to estimate how they would work in a given environment (they do not yet *know themselves*), nor do they know the workings of the world of science well enough to estimate how a particular type of employment would work for them (they do not yet *know the tradecraft*).[k] With this book we are furnishing tools to do both.

1.2 Match your goals to your character and talents

Which of your character traits are likely to lead to success? Sufficient basic scientific talent (including of course intelligence) is obviously essential to be successful as a scientist. While raw talent can be refined, it cannot be created. One tragedy that is common (but not always recognized) occurs when the talent level required for success in a particular field exceeds the raw talent of the scientist. In effect the scientist has, so to speak "run out of talent." The only real cure is to move into a different field. However, before doing this, provided the will is there, talent can be made much more effective by making best use of some particular traits and also by developing and strengthening others. In effect, "Do not quit until you have given it your best shot."

It is clear that, other things being equal, one would expect success in accomplishing science to be positively correlated with this native raw ability. While one can consciously try to explore various ways to give this ability free rein, the basic ability itself is probably not something that can be consciously learned or developed. To find one's basic level in any field it is vital to have a just opinion of one's basic ability in that field. This can be very difficult to do, and, while it is not something we will discuss here, the beginning scientist should make a serious attempt to do this.

---

[k]Another cartoon of Sidney Harris has a pensive garage mechanic saying to a customer, "Actually I started out in quantum mechanics, but somewhere along the way I took a wrong turn." (as shown on p. 141 of his *Chalk Up Another One*, Rutgers University Press (1992)).

Fortunately, and contrary to most people's beliefs, it is not the raw talent for science that is *the* most important trait that ultimately determines a researcher's success. It is little realized how much success in science (and in particular in experimental science) depends on other behavior which, unlike basic raw ability, can be learned, improved and developed. This book is aimed at going through what can be improved and developed, so our readers can decide for themselves how to be as good a scientist as they can. Development in this way is a means to the end just as much as the other kinds of research support. Another analogy is sailboat racing. There is a basic ability or "touch" in being able to coax a little more speed out of a sailboat, but there are also many things than can be consciously learned, anticipating tactical problems, boat preparation, weather prediction, many aspects of sail trim etc. There are books which describe what to learn about every thing except the sailing "touch."[1] This book on Survival Skills in Science is aimed at telling you about everything except the "touch" in your field of science.

To drive this point home, let us make it in another way. There are many admittedly clever and ingenious research scientists who are more talented than most others but who are not as successful as you would expect them to be, based purely on their talent. Certain character traits, such as drive, patience, and the ability to work in a team or to lead one, are extremely useful and perhaps more important, in the long run, than raw talent. Luckily, for those who learn how to do it, many of these other important traits can be consciously learned and strengthened. Of course you have to know which of these traits are worth cultivating and strengthening, and how to do it. Potentially brilliant concert pianists who cannot school themselves to the discipline of relentless practice, will remain just that "potentially brilliant," while those with somewhat less talent but more discipline can fulfill all their promise. (We will return to this topic when we discuss actions which will follow the basic choices.)

---

[1] Curiously enough, the sailors with the "touch" are just as obsessive about boat preparation as those of us who are doing it because we are obsessed with catching up to them. It seems that although "touch" is worth far more than sandpaper hours, the touch arises in part from being very sensitive to boat performance, so the "touch" sailors are almost in pain on a poor boat. The best sailors are thus unable to forego the sanding — they are too obsessed. The analogy with science is all too evident.

In the meantime, it is obvious that, given your particular character traits, there are certainly some fields or modes of working that are more suited to your character than others. A rational way to handle this is to decide what "research style" is likely to appeal to you, since that is probably the best way to channel your energies.

1.3 Work style choices: Lone wolf, collaborator, team player, team leader? Alpha scientist or beta scientist?

1.3.1 *What kind of teams operate in your science?*

One of the aspects that is important to determine is the nature of the normal method of functioning in the domain of science to which you feel called. The basic concept is the size of the typical team in the field, and this can be estimated by looking at the number of authors on a typical excellent paper, and by looking also at the author affiliations.

DIVERSION By the way, since we are talking about aggregations of scientists, it seems appropriate to indicate from the *Journal of Irreproducible Results* v. 14, n. 4 (1965)(ANON) some collective names in basic science:

A pile of nuclear physicists, a set of pure mathematicians,
a field of theoretical physicists,
an amalgamation of metallurgists, a line of spectroscopists,
a coagulation of colloid chemists,
a galaxy of cosmologists, a cloud of theoretical meteorologists,
a shower of applied meteorologists, a litter of geneticists,
a batch of fermentation chemists, a colony of bacteriologists,
a wing of ornithologists, a complex of psychologists.

One also finds more "political" varieties: an intrigue of council members, a dissonance of faculty members.

Something like five authors or less indicates a typical small team with, typically, a team leader or dominant scientist (we will be calling such people alpha scientists; they are not usually the first author), a student (or two or three, one of whom is likely to be first author), a post-doc (or two), perhaps an "intermediate" scientist (what we term here a beta scientist) and sometimes another collaborating alpha scientist, likely from another institution.

As one goes up to something like ten or more authors, the work is likely to be the result of a coalescence of smaller teams working on different aspects of the research.

Completely out of the range of these numbers are the papers on high-energy particle physics. The author lists here can reach several hundred, where the likely analogy is a military regiment. In the psychology of groups (which runs from something like three to about eighteen individuals) we are dealing with small groups, large groups and large aggregations undoubtedly composed of alliances between groups.

In a letter to *Physics Today* (November 1964) Robert A. Myers drew to the readers' notice that "Once again (F. Bulos et al., *Phys. Rev. Letters* v.13, p. 486 (1964)) the high energy physicists have been presented with a paper that has more authors (27) than paragraphs (12). Can high energy really be so different?" High energy physics is indeed that different, and now even more so. In a recent paper (Abe et al. (192 authors in the Belle collaboration) in *Phys. Rev. Letters* v.95 p. 101801 (2005) produced fewer paragraphs (18) than the number of institutions involved (49).

Yet another mode is that of a scientist (perhaps with two or three local collaborators) who carries on extended collaborations with other groups as a roving specialist collaborator or sub-group. (This is a mode which one of the authors (Tudor Johnston) has employed for years with considerable success). You should talk to people in the groups pertinent to your science and find out how the research is carried on and compare it with what goes on wherever you happen to be. Then you can make a rational plan, which may include moving your domain of research somewhat, to be able to operate in an environment more suited to your preferences.

In answering questions of this type, you should try to assess your abilities to work with others as objectively as possible, and then see to what extent you are willing to compromise. You should not necessarily view this kind of compromise as something negative. It is very likely that you will have to live with this choice for most, if not all, of your career. It is true that at the beginning one often has little choice but to be either a sort of laboratory technician/student or a low-level team player, the possibilities for real choice emerging only somewhat later. Nonetheless you should have clear in your own mind which way you want to go well before the first opportunity arises to make your own choice.

Once you have analyzed the field of interest to you, the next step is to think carefully of what style of scientist and science would make you happiest.

1.3.2   *What sort of player are you going to be in the game of science?*

Let us begin with an obvious analogy, namely team sports and the role of the individuals on the team. In team sports, the players have their own specific role. Take soccer for example. There are eleven players on the field: typically one goalkeeper, four defenders, four midfielders, and two attackers. In a scientific career the situation is very similar, except that we maintain that there are fewer roles which we are to call alpha, beta, gamma and specialist collaborator.

In this terminology, an alpha is a scientist who likes to think creatively and to transform his thoughts into funding for his/her research. Typically, that is the role of a professor in an academic setting with a significant research output. Often enough, an alpha does not have time to spend hours in the lab or to do the actual calculations (and may not be interested). An alpha will rather coordinate and *manage* a group of students and/or post-docs who will perform the experiments or simulations on the current concepts as seen by the alpha and the current concepts on behalf of the alpha.

This state of affairs has the advantage of allowing the alpha to pursue multiple projects in parallel, by delegating them to individual members (betas, most likely, or alphas in the making) of the team. The disadvantage is that as time passes, you (as an alpha) become more of a manager and less of a scientist, and eventually may lose contact with the laboratory (which is probably the reason you decided to become a scientist in the first place). The larger the group the more developed this trend.

A beta scientist, on the other hand, is more like an executive officer in the navy. The beta likes to spend most of the time in the lab, helping students with their experiments or even actually turning the knobs. Someone who gets things done! Occasionally a beta will help with "administrative" tasks, such as drafting grant proposals. However, a beta's heart is in the lab, or anyway a beta strongly prefers turning the knobs, doing the calculations, and so forth. This sort of figure is extremely precious, because a beta will help supervise graduate students and post-docs and will make sure that things are running smoothly while the alpha is away, teaching or otherwise busy.

The difference between an alpha and a beta can also be compared to the difference between the captain of a ship, and its executive officer or "exec". The captain is in charge of the ship, and of its overall military strategy. The ultimate responsibility for failure or success rests on his shoulders. The exec is second in command, and is there to execute the captain's will and orders, or to make sure that the sailors execute properly.

Below the beta lies the gamma scientist, who does not really want any executive responsibility and is happy to be absorbed in a particular specialty, with the occasional publication (as lead author) on the nuts and bolts of the cherished sub-system in (say) the *Review of Scientific Instruments*.

The decision of whether you will aim at being an alpha or a beta or a gamma is another critical one (just like deciding whether you want to be an experimentalist or a theoretician). You should analyze critically and *coldly* your skills and personality traits, and determine as objectively as possible if your profile better matches an alpha or a beta or a gamma. If you are in doubt, we advise you to present this issue to colleagues/friends and get their feedback. You may also want to ask your supervisor, if you are a graduate student. Make sure they understand this is a sensitive point, and that they should give you an objective answer. Scientists are all born as betas, and as long as you work for a supervisor, you will always be a beta at best. The question is whether your advisor sees you as a potential leader, i.e. as a future number one or alpha, or as a beta.

In our opinion, this is an important issue that will strongly affect your career — positively if you choose wisely, negatively otherwise. Imagine a soccer player who is good as a goalkeeper, but desperately wants to play as a midfield or a central forward. Or a centre-forward who would rather play defense. It will simply not work. Although the analogy cannot perhaps be translated 100%, the general idea is clear. Do not try to be someone you are not: it would be disastrous. Play in your role and you will have a much better chance to be happy and even to succeed.

The North American academic system is designed to host and promote alphas almost exclusively — at least for tenured positions (the few that are available). In a university in North America, a beta in a small group would be called a *"Research Associate,"* but in a larger group might have a more respectable sounding position such as "research professor." These are very respectable positions, and if you decide this is your natural role, there is nothing wrong in seeking this type of job (even though the title of "professor" probably sounds more prestigious). However, you should be aware that — with few exceptions — the salary of a Research Associate typically derives from *"soft money."* The position is renewed as long as there is an alpha to bring in funds and grants. It is almost impossible to turn it into a permanent position.

In Europe, on the other hand, the academic system is more hierarchical, and groups of professors often work in teams, led by one professor at the top. In such cases, not all team members can be alphas (for obvious reasons — they would be stepping on each other's toes all the time), and thus the system allows (and to some extent favors) hiring a good number of betas into permanent faculty positions.

Jobs in industry and government labs, by their very nature, are more appropriate for betas. In these environments, the management will typically set the objectives and tell you what to do on a short term, medium term and long term basis, and you will be the one turning the knobs and executing the research, first hand. Normally there will be no graduate students to delegate to, but occasionally there can be post-docs (depending on the philosophy of your employer, among other things). In these settings, being an alpha means becoming a manager, and probably doing less and less in terms of research (although again, it is difficult to generalize).

Federico: — Over the years, I had many lively discussions on these aspects with an astronomer friend of mine, who really likes to do research but cannot see himself in a faculty position. He dislikes the idea of teaching (perhaps he would consider a little of it at most) and of writing grant proposals. When I pointed out that as a professor he could advance his ideas and projects faster, by working with several graduate students in parallel, M. replied: "*I want to DO research, not manage it.*" I think this sentence, although a bit severe, more or less summarizes the key differences between an alpha (who now manages research, and does little of it himself) and a beta (who actually does it).

1.4   Choices in work climate

In general, doing science and research requires teamwork and coordination (among other things). If the work is done in a team format with a quasi-military hierarchy, there are many roles, such as supreme leader, group leader, team player, outside specialist and (temporary) slave labor (i.e., graduate students). While most people prefer telling other people what to do than being told, the road to the top job can be arduous, and the leader's job may well become more like the president of a small company than a scientist. If the leader's job seems too political, being a group leader may be a useful compromise. More often in science the work is done in collaborations which are much looser and more democratic, probably less stressful with more room for individual roles in a pleasant group. Things run best when everyone plays a role in which they are at ease. Unfortunately it seems to be the case that the North American (academic) system tends to evaluate everyone as if they were trying to be number ones. Even if Napoleon said that every soldier had the baton of a field marshal in his knapsack, he only meant that the top job was open to anyone with talent, and not that sergeants should be judged by their ability to command armies. The European system privileges teamwork and has room for everyone, but North American criteria for promotion and the like often seem to place too much emphasis on the scientist as individual entrepreneur.

1.5 A basic choice: Experimentalist or theoretician?

Although this is an aspect that applies mostly to physics and to chemistry (in the sense that it is a basic choice for those fields), there are aspects that apply to other fields, in which one does find a cultural divide between the experimentalists who produce the data of what might be called the "ground truth" (by analogy with aerial reconnaissance), the touchstone by which all theory constructs are tested and the people who create the theories.

In physics theorists are very well-known, but in other sciences the theorists are less prominent. Probably the most prominent chemistry theorist in the 20th century was Linus Pauling (Nobelist for the work discussed in "The Nature of the Chemical Bond"). In biochemistry, the best-known examples of what we here term "theoreticians" are the members of the Watson-Crick duo who created the DNA model concept.

DIVERSION Theory and theoreticians are hard to represent, but there is a glorious counter-example, shown on the next page. The actual object of the cartoon is an *equation*! (Guess which equation it might be before you turn to the next page.) The cartoon is by Sidney Harris[1] (naturally!) and is the only one (to our knowledge) where an equation from theory is the real point, but one where most non-scientists can get the joke. (Permission for use of this Sidney Harris cartoon was received via his web site at ScienceCartoonsPlus.com.)

Federico: — As an undergraduate student, I was particularly good at theoretical coursework, and was hoping to become a theoretician as a career. (In several Mediterranean countries (perhaps for cultural or historical reasons) doing theory is considered more popular and prestigious than getting your hands dirty with experimental work. In Italy, most physics students begin by hoping to become theoretical physicists.) This early inclination is also related to the fact that I was (and still am) somewhat clumsy in the laboratory.

As time progressed, however, I came to realize that all the best students of my course (and many of them were better than I, at least when comparing primitive indicators of performance such as grades) wanted to become theoreticians as well. Not only, they all wanted to pursue the theory of high T_c superconductivity, a topic in condensed matter physics that was — and still is — very hot and controversial at the time, and which I was also interested in. Moreover in the physical sciences there is often the (unfortunate) perception that doing theory is to be considered more noble and prestigious than getting your hands dirty in the laboratory.

Thus I was faced with a very common kind of dilemma: should I follow my instinct, and become a theoretician, but accept being at the end of the list? Or should I make a compromise and choose to become an experimentalist, perhaps fighting a bit against my inclination, but probably being the best experimentalist in my course? I ended up choosing the latter. It turns out that it has helped my career enormously. Obviously at first it was not an easy choice, but in hindsight it was by far the best choice for me.

The trick to being happy in science, as with life in general, is to know yourself well enough to make the right choices, choices that you will probably have to live with and that may be irreversible. Incidentally, several of my theorist colleagues were also highly successful so far; however, in the face of tough competition, more than one gifted student who stayed in theory dropped out of graduate school and ended up pursuing a completely different career, like consulting or financial mathematics. This is not necessarily a bad thing of course, and you could argue that perhaps they were not meant to be scientists after all. They may even be happier doing what they are doing now. However, if they had made a different, more rational choice like the one I did, perhaps they would have stayed in science. (I know at least one person who wonders what life would have been like if she had stayed in scientific research, and she probably will ask herself that question for a long time hereafter.)

Tudor: — My anecdote is almost the reverse. I began as an engineer (actually a hybrid called engineering physics), did an essentially experiment-plus-theory-interpretation engineering thesis at Cambridge University, and began a mixture of theory and experiment in a Research Laboratory doing contract research for the government. As time went on I realized that there were what I found to be deep mysteries in the experimental technology — vacuum leaks, electrical hum, electrical ground loops, to name the worst — which I could never master and which remained for me "wild magic." I gradually but happily abandoned my feeble attempts to master the arcana of the laboratory and became a theoretician often for experiments in which I had no direct interest, becoming in fact a *de facto* theoretical physicist in plasma physics, although holding no physics degree.

One advantage of this long love affair with experiment is that I retain a deep respect for people who can make experiments work, a lively interest in how experiments are done and how to (in effect) "diagnose" any theoretical constructs to suggest experiments. It gives me a special thrill if I can use theory to show how an experiment that appears not to be working can be saved by using a different protocol and analysis procedure. A theoretician I have become indeed, but a better one for having been also an experimentalist for some ten years.

Now all this was accomplished with little conscious analysis on my part, without any real penalty, but from this I have become convinced that doing frequent "reality checks" on your comfort level in the way you function in your science is essential. Without this there is a significant risk of drifting into a way of working in which the basic mental discomfort of a poor fit between you and the way you are working leads to a dysfunction in the work for reasons that may pass unperceived.

Federico and Tudor: — To place things in perspective and make this example relevant to scientists in other fields like biology and chemistry, where theory as such may not bulk as large, we point out that each discipline has its "hot" subfield. In biology it might be molecular biology, and in chemistry something else.

In the anecdotes just reported we suggest a generalization with "hot subfield" replacing "theory."

The lesson to be drawn from the first anecdote thus becomes this. Do not go into the "hot" subfield of your discipline just because it is trendy and all the good students want to do the same. *You* have to choose what is hot for you (Know thyself!). Later, when you become a successful scientist you may even create a new trend and a new "hot" subfield. In the longer run, your goal is to become a leader, not a follower.

The lesson to be drawn from the second anecdote becomes the following.

Just because a field is "hot" is not a reason for going into it, and the competition will be fierce if you do "jump into the hot water". However, do go into the "hot" field *UNLESS* BOTH of the following apply: (i) it is sufficiently attractive to you for its intrinsic worth and (ii) you find that you are generating interesting ideas almost in spite of yourself.

Chapter 2

Basic Strategies and Actions

Sections of this Chapter

With your basic choices more or less set from the previous chapter, it is time to discuss strategies and basic actions in this chapter, with certain aspects to be amplified further in Chapters 3 through 5. By the way, although we have mentioned Feibelman's book because it was a major influence, there other books of a like nature,[3] and they are also well worth examination.

2.1 Career choices vs. personal choices and reconciling two professional careers

Before discussing individual choices, it is worth noting that the issues are more difficult to settle when both people in a couple are involved. Life is considerably more difficult if both partners in a family follow highly specialized vocations. It is rarely easy to find the right balance, or compromise between your careers and your private lives. Poor planning in relation to this issue can damage both. Although there are no easy answers, good planning usually helps.

The main question (and the hardest question to decide) is which of the two partners has priority when it comes to a particular career move. As for all difficult problems, it is better to at least discuss the problem in the abstract (and perhaps settle the terms of the discussion) long before the next move looms on the horizon. Even when decided once in a particular way, this priority is not settled forever, and may switch back and forth between the two partners as circumstances vary. Beyond this, circumstances vary so much from case to case that little more can be said, beyond repeating that it is better to speak of the problem in principle and try to settle the ground rules well before the actual event.

2.2 Choosing a University (Ph.D., Post-Doc, Faculty)

As usual, for most questions, the right answer for you to the question of "Which University?" depends on your circumstances at the time. Probably the question is least important when considering it in connection with a position as a post-doctoral fellow, because it is for a relatively short time (typically two years). When embarking on a Ph.D. at Prestige University, you presumably realize clearly that the "brand name" on your Ph.D. will have an effect on the reader of your CV throughout your professional life, long after you have left the hallowed halls of Prestige University. On the other hand, when considering, say, an entry-level tenure-track position or a research professor position, the actual circumstances weigh more heavily than the "brand name," since you may well hope to be there for a relatively long time, up to five or six years or more. Put in another way, for this case, the name of the university will then be important to you (hopefully) only if you choose to leave it.

> **DIVERSION** The Ph.D. is a fairly recent invention; it is an outgrowth of the German intellectual modes of the 19th century. ... Thus many eminent mathematicians of times past had no Ph.D. One day someone approached one of these august gentlemen and asked, "How is it that you have no Ph.D.?" The gentleman drew himself up and said, "Who would examine me?" From *A Mathematical Apocrypha*.[1]

The Ph.D. did not become established in Great Britain until the 1920's. Thus, for instance, G.H. Hardy did not have a Ph.D. His comment on the degree was, "A German invention, suitable for second-rate mathematicians and foreigners." (From Steven G. Krantz, p. 127 of A Mathematical Apocrypha,[1] published and distributed by the Mathematical Association of America (2002).)

This said, how should one choose a university? Should it be by reputation or "brand name"? Be careful! In North America, in particular, some universities are very famous one may say "super-famous", and being able to list them in your CV at any stage of your professional development would certainly be very advantageous for your career.

(Examples include Harvard, M.I.T., Cornell, Caltech, Stanford, UC Berkeley, Yale, and Princeton in the United States, and the University of Toronto, the University of British Columbia, and McGill University in Canada.)

However, although this is an important fact to keep in mind, this should not necessarily be your first criterion of choice. As will be discussed below, there are other important aspects that you should consider. In the big picture, you want to be happy, and choosing your employer only on the glittering grounds of prestige could mean that you are giving up a lot in terms of your personal life, as well as in terms of other aspects of your job and career. (Remember, *all that glitters is not gold*.) How to order these aspects in your priority list depends on how you weigh your career in science with respect to your personal life.

These aspects other than the university brand name include the city or town where you will live, the project(s) you will be asked to work on, whether you are facing a two-body problem (i.e. your partner is also a scientist or professional, and therefore would also need to get a position in the same city), and whether you are likely to get along with your prospective advisor or employer and the other members of their group. If entering a doctoral program, for something like up to five years, the choice of location as such is not terribly important, since you are likely to leave after your degree. However if you really dislike the place, it could become a problem and make your life miserable (also with the possibility of bad consequences on your scientific performance).

Of course if you are considering a possibly permanent stay (in connection, say, with a tenure-track position) these living conditions aspects loom much larger. (One aspect, not relevant for a short stay but for the very long term, is that often the private universities give very generous breaks to faculty members on their children's tuition.)

In smaller schools, at Friendly College, say, faculty members tend to be more "collegial," as convincingly described by H., who used to be a professor at a very famous Ivy League School before relocating to a State University in the South of the U.S. His professional life seems to have actually improved substantially in the aftermath of his move from Prestige University to Friendly College. For one thing, he is tenured now. (Most big schools rarely offer tenure. They tend to milk the best out of their junior professors when they are young, then get rid of them as if they were old shoes.) Even better, now H. can concentrate on his work without feeling the constant, unpleasant pressure of having to produce outstanding results, and of bringing in ever more research funds. Finally, he was recently promoted to Full Professor.

Of course, we would all like to continuously churn out fantastic data, and to have fat research accounts. However, having to do it under conditions of duress and perpetual stress can make it all the more difficult, even unpleasant, and actually hamper your progress.

On the other hand, residents of famous Universities will argue that they have the rare privilege of working in a very "special" place. Since such universities tend to be extremely selective, and since such workplaces are indeed considered so prestigious, there is an obvious element of truth in their statements. But it is still true that the benefits of the reputation of Prestige University become really important on the CV that you will use only if you leave. Professor H. was certainly helped by the reputation of Prestige University when applying to Friendly College.

As we remarked at the outset, the tradition and record of scholarship and accomplishments of the University where you carry out your graduate studies and your post-doctoral work will have a strong and lasting impact on your personal and scientific growth. Once again, you will have to determine what is best for you and which environment will offer you the highest chances of success. These stints however tend to be more limited in time, up to a few years, so in some sense the choice of location for this purpose will not have a *permanent* effect on your private life.

In general, when applying for a tenure-track or research position above the post-doctoral level, you should beware of scientists who want to employ you as a form of cheap labor, even if that is what you really are from a — strictly — "market" point of view. Even if you get an offer from Dr. Famous, Chair of Department X at Prestige University, you should make sure that this prospective employer will be interested in you and respect you as junior faculty, and not simply treat you as a super-post-doc who does the work and gets the papers published. To determine this (if you can make the contact), the best source is a recent post-doc no longer with the group, the next best source being a current member of the group.

If you decide to go for a post-doctoral experience, as nearly everyone has to in order to begin to establish a record of research independence, you should plan to make it last *at least* two years.

To set the scene, we remind you here that the cheap labor aspect discussed in the previous paragraph is the foundation-stone of the post-doctoral fellow system. As so clearly explained by Feibelman,[2] Dr. Legree's object is to get the maximum out of you in the two years or so of your time with him, while your object is of course to add to your CV and (we hope) to learn some new science and technology, but also to bounce to another and better position. If you are good and lucky, you may indeed find a better position before the two years expire. However, in general, two years is the average time it will take to acquire new skills, get new results, to publish them, and to place yourself successfully on the job market. If you plan at the outset to stay for a shorter time, you may find it hard to get a better position. Of course, you can choose to move into another post-doctoral slot. However, in a first post-doctoral stint, this is an option to save a situation gone sour. Ideally the one post-doctoral experience of two years should be enough to get a semi-permanent position afterwards, be it in industry, a government laboratory, or a faculty position. (Clearly, for the reasons we just outlined, one single two-year post-doc is far better than two one-year post-docs; which will leave a flavor of frenzied desperation for that period on your CV.)

Given the level of competition, many find it preferable to go on to a second post-doctoral position if an acceptable tenure-track position or something comparable has not been offered. Much the same remarks apply as for the first, except that now taking a significant leap outside your previous area of research is something to consider seriously, to broaden your target area of employment and to demonstrate your versatility. A third post-doctoral stint smacks of desperation or of excessively narrow standards and is to be avoided if at all possible.

2.3   Why go through the post-doctoral apprenticeship?

One of Federico's students once asked him, "What is the use of "*going through*" a post-doctoral apprenticeship?" Federico replied that while unsure of the origin of this temporary form of employment, he suspected that it had originally been invented to conveniently exploit people. However it also answered a need for the young candidates in providing a way of staying in their field if there was a shortage of more permanent positions. It soon became a necessity for the fresh Ph.D.'s, since it is hard to compete for a faculty position straight out of your Ph.D., when all the other applicants have gone through at least one if not more post-doctoral jobs with the additional strength that the extra experience brings (their publication list tends to be much longer for obvious reasons).

In the end, post-doctoral experience is useful not only because it usually increases your number of publications, but also because it should represent the experience that turns you into a truly independent scientist, in effect putting the finishing touches on turning you into a scientist capable of independent research.

You are not likely to be mature enough to lead your own group when you have just finished your Ph.D. A doctorate should signify that you are able to carry out a project on your own, or even several projects. It should mean you can plan experiments or calculations, acquire data, analyze it, and finally write it up and publish it. However, it typically does not give you enough maturity to lead a group or build up a successful research program. Most scientists have to go through more professional development and training before they can function independently.

During a visit to his lab, a colleague (and friend) described scientific training to me (Federico) in the following terms. "On average, and by and large (which means that there may be exceptions to this picture) when you are a Master's student, 90% of the time you are doing what your supervisor asks you to do, and the remaining 10% you are giving a personal and original contribution. On the other hand, if you are a Ph.D. student, on average your personal contribution will be of the order of 40 or even 50%. You will be expected to exercise leadership in several aspects of your project, and to take responsibility for your decisions and course of action. Finally, when you become a post-doctoral research associate, your expected contribution to the project will be of the order of 90% or more, and in fact you may be asked to train and supervise other graduate students during your tenure." I tend to agree almost completely with this analysis.

All in all, a post-doctoral experience will expose you to different aspects of a scientific career, which you would normally not go through during your graduate work, and will give you the opportunity to lead a project all by yourself. It should also be a time of great productivity, since you can devote yourself pretty much full time to research and do not have teaching assignments and are not expected to spend too much of your time writing grant proposals (which is what happens instead when you become a faculty member). In essence, it is an intermediate period which should give you enough time to reach full maturity before you start your own research group. It may also be your last chance before deciding that after all a scientific career is not what you want to do with your life. In a more positive light, it can help you choose between the three types of career, industry, government or academia. There are, of course, other possibilities. Several people I (Federico) know did an MBA after their post-doctoral experience, and then either started their own company or became consultants.

In looking for a post-doctoral job, many (perhaps most) students want to pursue a new set of projects, which will enrich their overall scientific background and professional development, hopefully also learning new techniques and approaches. They typically look for an advisor who has a solid reputation, possibly a nice personality, and a wide network of national and international connections.

We deliberately avoided making any point about the choice of science you want to pursue in your postgraduate work, but when it comes to postgraduate work there are some new points to consider. Fundamentally there is a choice of whether to go on exploiting the narrow expertise you have acquired or whether to try and make a "lateral arabesque" to demonstrate that you can learn a new field quickly and broaden your possibilities for positions.

Preference is often given to activities which extend and complement what the student did in their Ph.D. For fairly obvious reasons, this is what we call the *"follow your own crowd strategy,"* since this is what most aspiring scientists want or try to do. There is nothing wrong with this rather conservative strategy, which is a fairly good default choice. However, since most students will choose their post-doctoral job along these general criteria, once the experience is finished they will find themselves competing for a job. Since they will all have developed essentially the same credentials and skills as the rest of the crowd they will all be competing for the same general pool of jobs with deepened but not broadened expertise.

A contrarian's suggestion would be to break out from your current crowd, by choosing an ambitious, potentially high risk/high payoff project which is not on the track you and your competitors are currently pursuing. By this we mean trying to identify an area of science (naturally not enormously far from your general technical capabilities) which is presently underdeveloped but which is yet emerging and growing rapidly.

If your work during this more daring kind of post-doctoral experience is fairly successful, you will have contributed as a pioneer in this specific area, and this is very likely to give a powerful boost to your career. As a bonus you will have also demonstrated your ability to learn new science quickly. In fact you may be immediately considered one of the early leaders, and so, as more scientists start working on this topic, you will receive more and more recognition. Under these circumstances, rather than competing with hundreds of applicants for the same position(s), as in the conventional strategy, you will have a tremendous edge on your erstwhile competitors and in fact jobs will probably start looking for you rather than the opposite.

If you find the risk unacceptable when considering your first post-doctoral position, you should seriously consider it in a second post-doctoral effort (not a rare occurrence), after a more conventional first effort along the lines of your doctoral work.

The chief practical difficulty in following this less conventional choice is that you will have to make a bet on what particular area of science to choose. The odds of making a good choice may well be against you, since the information at your disposal will be incomplete. You may end up choosing a topic which is indeed underdeveloped but one which may not explode satisfactorily during your post-doctoral experience, either because it is not ripe enough or because it may have just looked more promising than what it really was. Unfortunately, there is no perfect solution to this. All you can do is search thoroughly and then take your chances.

Because this scenario has more risk, you should find an advisor who, in addition to the qualities described above, is willing to let you spend a certain fraction of your time on less risky "side projects," for example co-supervising and collaborating with graduate students who are already into their work. This has the added benefit that you will further develop your teamwork skills, learn something about group management, and also keep publishing in case your high risk project does not pay off.

What you should absolutely avoid, (both as a graduate student and as a post-doc) is to begin working in an area of science which is already saturated. This could well be "scientific suicide." You should only consider working in a saturated field *if you are convinced that you have what it takes to solve its main problems.* Without this assurance you are joining this large mass of people, with whom you will be competing for funds, positions and general resources all the way through (at least until you start working in a different area).

A personal anecdote may help to clarify this. When I (Federico) was an undergraduate student, high T_c superconductors were all the rage for condensed matter physicists. I thought it was an interesting problem, and considered it seriously for my graduate work. However during one class, a professor pointed out that, since the discovery of the phenomenon less than ten years before, tens of thousands of papers had been published on that topic, and yet nobody was even close to understanding the problem and being able to describe the experimental results. That was enough to convince me that it would be unwise at best to enter that field. The situation does not seem to have improved much since then, so I am sure that my decision was definitely the right one.

2.4   First Career choice: Thesis advisor — Young or senior?

Some people will say that it is almost always unwise to work for a young professor, who will be to some extent competing with you for credit. (A young advisor is naturally hungry for credit since there is a time of only about six years to demonstrate enough productivity to become (hopefully) tenured. In P.J. Feibelman's book, *A Ph.D. Is Not Enough,* you will get that impression.) Of course a young professor also needs to attract good students and post-docs to the lab. Indeed if this notion were considered by all to be 100% true, it would be next to impossible for a young faculty member to attract students and post-docs, in which case it would hardly make sense for a department to hire at the junior level.

However in most cases, if the young professor is very talented, he will not regard you as competition and all will be well. The risk arises when the young would-be thesis advisor is struggling and unsure. Now you see the necessity of obtaining an accurate picture of ability and character of the thesis advisor before you are fully committed. While it always helps to get advice from other graduate students as to his behavior as an advisor, getting a good estimate of his ability can be a little more difficult. You should begin with a study of the advisor's CV (if this is not forthcoming you should begin to suspect a problem). Clearly, if a young professor already has tenure, then you have the advantages of a certain guarantee of competence (even brilliance) and of approachability without the worry of competition from one's thesis advisor.

One can hope that senior professors are more likely to view their students as their research children, at least while the advisor-student relationship is maintained. (This point is made by Feibelman.) However, if a student is particularly bright, and finds employment elsewhere, there may come a point, particularly if the ex-student becomes very successful, when the ex-student is seen as (ungrateful) competition with the former advisor. ("How ungrateful it is of my ex-student to compete against me, particularly in the field in which everything was learned from me!") To some extent this has happened to one of us, and we can cite other examples of this; it is not as uncommon as one may think.

In the end, choosing the right advisor is very difficult, and although a senior advisor may offer some advantages, this is not always the case. A younger advisor is more likely to be enthusiastic, energetic and dynamic (and perhaps more naïve at the same time). A more senior professor, though perhaps less creative and dynamic, will have a lot of experience, and will therefore be in an excellent position to give good advice. The mature advisor will also have many connections in the community, which will undoubtedly be useful when you finish and are looking for a job. Of course, luck will play an important role in the choice of a thesis advisor, just as it plays a role in everyday life. You want to reduce the element of chance to a minimum, but you cannot eliminate it altogether.

When you meet a prospective advisor, you should not be afraid of asking for a copy of their CV. From this Curriculum Vitae, you will learn a lot about this person. You will be able to see how much has been published (you should take note as to whether the student's name is put first), and perhaps other important aspects, such as their funding situation. More and more professors have established Web sites and these can be excellent information sources as well. You should find out how many students this scientist has in hand, and possibly try to talk to them. (Here a reluctance to have you talk to the students is not a good sign.) Even better, you should talk to this scientist's "alumni," not only to learn about their experience, but also to see how they did after graduating. Was it difficult to find good employment? Did the training offered by Prof. Seldom Available prove to be useful? Or was Professor Mediocre so ill-respected (albeit highly available) that those students found his reputation almost a hindrance? University research is all about training, and your goal is not just to check the scientist's science credentials but to determine whether you are likely to obtain good training there. With all this information gathered, now is the time to make a choice.

Overall, in choosing, you should think of and try to balance several aspects such as, for example:

(i) Do you like the project which you are going to be working on? Working for Dr. Famous may be great, but if the project that is assigned to you appears not to be interesting, you are better off changing project, if not also the advisor. If you want to succeed, it is important that you like your project, not only because you will be responsible for it, but also because you must be willing to work very hard for it. Who would be willing to work hard on something they do not really like?

(ii) Do you get along with Dr. Famous? Do you think you will get along once your project is finished and you are ready to move on to your next job? This is not a silly question. Your advisor will probably have to write letters of recommendation on your behalf for several years, if not more, long after you leave their lab. Therefore getting along with this person will have a major impact on your career, not just now but also well after you leave. This may be another imperfection of the system, but once again, it is reality, and you should be aware of it. Be honest with yourself about whether you will enjoy working in Dr. Famous' group. After all, if the relationship does not work out, you are the person who will be more damaged, who stands to lose the most. Know yourself, and try to know your neighbor (including Dr. Famous!).

DIVERSION   A little story on thesis advisors is available on the internet.from: north#NoSpam.hgl.signaal.nl (S.North).

In a forest a fox bumps into a little rabbit, and says, "Hi, junior, what are you up to?" "I'm writing a dissertation on how rabbits eat foxes," said the rabbit. "Come now, friend rabbit, you know that's impossible!" "Well, follow me and I'll show you." They both go into the rabbit's dwelling and after a while the rabbit emerges with a satisfied expression on his face.

Along comes a wolf. "Hello, what are we doing these days?" "I'm writing the second chapter of my thesis, on how rabbits devour wolves." "Are you crazy? Where is your academic honesty?" "Come with me and I'll show you." ... As before, the rabbit comes out with a satisfied look on his face and this time he has a diploma in his paw.

The camera pans back and into the rabbit's cave and, as everybody should have guessed by now, we see an enormous mean-looking lion sitting next to the bloody and furry remains of the wolf and the fox.

The moral of this story is: It's not the contents of your thesis that are important — *it's your Ph.D. advisor that counts!*

There is of course no such thing as an ideal supervisor who will be so for every student. You may find an advisor who is ideal for you, but that same person may be quite the wrong advisor for another student or post-doc. By the same reasoning, while the fact that his students get along well with Dr. Famous is a good sign, it is an indicator (and not a guarantee) that it will work for you.

(iii) Do you think you will like living in the city where Dr. Famous' University is located? During your (supposedly and hopefully) brief tenure in Dr. Famous' group, you will spend long hours in the lab, of course, but you will want to have something interesting to do during your (little) free time. Your location perhaps should not be at the very top of your priority list (although I know many people who would love to live in Hawaii for example), also because your Ph.D. will last a few years and your post-doc even less, and you can certainly relocate elsewhere once you are through (actually, you will be encouraged to relocate elsewhere). However, your location should not be at the very bottom of your priority list, either.

2.5   Find a mentor as soon as possible

While working under a thesis advisor it may be a bit delicate to be consulting someone else on your career (i.e., using a *mentor*). However, if you can find a mentor that early, by all means do it. In any case, once you have emerged from the thesis advisor-student relationship, it is an excellent idea to find a disinterested mentor in your new position with whom you can discuss your career moves and from whom you can ask advice. Senior faculty members can be great advisors and great mentors, however you should not discard *a priori* choosing a younger professor as your mentor. Indeed a younger professor may compensate his lack of experience with his energy and enthusiasm (if you choose well). (There is no law about not having more than one mentor at a time.)

Finding a more experienced scientist who is willing to give you advice may be very helpful and may save you from making bad choices, and therefore a lot of grief. If you know yourself well enough and if you believe in yourself to the point of pursuing a scientific career, we advise you to find a mentor as soon as possible (if you are not apt for a scientific career, it does not make much sense to look for mentorship). The best possible mentor is someone with more experience than you, who takes interest in what you do and in the possibility that you survive scientifically, and who has nothing to lose or to gain from your success (or failure). In other words, a good mentor is someone who can provide an objective measure of reality, especially when you are faced with a critical choice.

In the rest of this chapter many actions will be discussed. If you are thinking of taking some decisive and proactive action, here is where having a mentor will be particularly valuable, especially helping you to decide if this is the case where, say, a battle should be fought or is it rather a case for patience. As in playing bridge (arguably the deepest card game), one may learn many tactics, but the deepest art is knowing when to embark on which line of play, the diagnosis, so to speak. Here is where a mentor can really shine.[a]

Consider the examples we will give later in Chapter 6 *Cautionary Tales* (i.e., real life examples), if T. (Sec. 6.3) and M. (Sec. 6.4) had been properly mentored, perhaps they would not have abandoned their scientific careers.

2.6   Choosing collaborators

While the early scientists like Newton and Galileo used to work mostly by themselves, nowadays there is a tendency to work collaboratively in pairs or even in larger groups. This trend has been largely caused by the size and complexity of modern research projects, particularly in experiment, which now requires grouping researchers (or even teams of researchers) with complementary expertise.

In pursuing collaborations (which is something we strongly recommend, since we are convinced that the most significant scientific advances of our times are the result of a collaborative effort), how should you choose your partners? Of course, the basic prerequisite is that a collaborator should have the same level of interest (and possibly of expertise, but in a complementary sense) on the subject to be investigated. This is, however, a necessary, but not sufficient condition for the partnership to work out effectively.

The most important personality trait of a good collaborator is that he/she must be *reliable*. Many scientists have a tendency to try to collaborate with *famous* researchers, on the assumption that they will be more respected by the community (*he works with so and so, who is very well known, therefore he must be good*). Unfortunately, famous scientists

---

[a]Federico: from this point of view, I must admit that I was particularly lucky, since I was mentored by my father since the very beginning of my higher education. Few others will be that lucky, and therefore will have to look for mentorship elsewhere.

are often extremely busy, and as a result they are very stingy with their time. They will thus accept working with you only if they perceive that they have something useful to gain from the partnership, and will often not be shy about reminding you that they are doing you a favor by accepting to collaborate with you. This can easily place you in an uncomfortable position, whether the collaboration turns out to be successful or not.

A "normal," *reliable* collaborator who is willing to invest time and effort in working with you on a regular basis is a far better choice. This collaborator may not be as glamorous as Dr. Famous, but that continuous investment will be of greater help in the long run than the odd few minutes of Dr. Famous' attention and his name on the grant or on a joint publication. The start may be slower, but over a medium/longer term it will give you a lot more satisfaction.

In fact, you should also be aware that if you publish together with Dr. Famous, you may end up getting almost none of the credit anyway, on the assumption that his contribution to the work in question was a lot more significant (after all, there must be a reason why he is already famous and you are not).[b]

You should make a strong effort to take a serious interest (and do your best to understand) all the projects of the members of the group (even if you are not the leader!) and also to take an interest in the neighboring groups. All scientists are generally kindly disposed to people who display their good taste and judgment in wanting to know more about their work. These are the people who will be part of your network in the future, and they will often be in a position to help you on your way up.

---

[b]This is sometimes called the "Mathew effect." From the New Testament Book of Mathew (ch 13, v 12) (King James version) we have, "... unto him that hath shall be given, and he shall have more abundance: but whosoever hath not: from him shall be taken away even that which he hath." N. David Mermin discussed this a bit in *Physics Today*, v. 57, May, pp 10-11, (2004) (letter comments in the v. 58, Jan, pp 15-16, (2005)) Merton, R.K. (He decided that he had Mathewed" himself, in attributing something he originated to Richard Feynman.) The usual prime reference is to R. K. Merton, "The Mathew Effect in Science," *Science* **159**(3810):56-63 (1968), but Louis F. Feiser and Mary Feiser were much earlier in *Inorganic Chemistry*, D.C. Heath (1944).

Whenever you set up a collaborative project, it is advisable to clarify each scientist's role from the very beginning. This will avoid unpleasant arguments later, if one of the collaborators has the tendency to do too much, or too little. Without being petty about it, make sure that everybody understands how the collaboration is going to proceed, and who is responsible for overseeing the whole project. Especially if the work requires a large number of collaborators, it is highly desirable to have a good project manager. This person should be a good organizer, and should be willing to take the overall responsibility for the success (or failure) of the project. You should avail yourself of every opportunity to improve your management skills, even if you are not a person with managerial responsibility.

2.7 Which character traits lead to success?

Many of you will think that *intelligence* (by which is meant *SCIENTIFIC intelligence,* or perhaps *scientific brilliance*) is the most important personality trait required to be a successful scientist. Indeed, there is no doubt that intelligence is a very important quality. However, we beg to differ somewhat: Yes, intelligence is essential to succeed. No, it is certainly not the only important aspect of a scientist's personality, and perhaps it is not even the most important one. We know more than one highly successful scientist who is definitely not as bright as many others. One can have an alpha career in science even if your science talent is no better than beta. What you have to do is to do everything else right.

So how does a basically "beta"-ability scientist become a successful "alpha" scientist?

One obvious component of success is to work very hard when you work, and also to work long hours. Another way is to make an effort to contribute well to several phases of the major project and to work at being a consensus-builder. (Consensus-building and being involved in several aspects can lead to becoming a spokesperson, a good step towards becoming a manager.)

Two of the most important personality traits very useful in these activities are strong personal drive and being able to work successfully in a team — especially if you are an experimentalist. (These two tendencies can often conflict, and skill in reconciling them in yourself is part of the "know thyself" mantra.) Another useful trait is being available and reliable. As you proceed in your career, the skills of being able to lead and to coordinate a team will also become very important.

However many of these traits for success are so deeply rooted in your personality that it is not realistic to consider developing if they are totally absent. One way to find out is to try the related activities (e.g., step up to manage a modest sub-project) and see if they suit you. In the end, what you have to do is to evaluate your personality and decide to what extent you can operate in these various modes. There are some useful traits, pertaining directly to your own work habits and organization, that can be developed and are always well worth the effort, as discussed next.

2.8   Character traits which can be developed: Self-organization, rigour in science and meeting deadlines

2.8.1   Be organized

Scrupulous organization of both your scientific work and your career development makes success generally very much more likely. For the scientist whose work may become important (and we all hope that our work will be worth looking at in detail), keeping track of everything (and being able to find it again years later) is one of the vital keys to practicing scientific rigor. Developing a reputation that your results are always reproducible by others (and if not it is the others who are making errors) is an absolute must for becoming highly respected as an experimentalist.

Developing a reputation for publishing results which others cannot reproduce is like having bad breath, everybody knows but nobody will tell you. People will instead quietly give up trying to build on your results after they have put in wasted effort trying to reproduce earlier results of yours.

This is another reason for publishing everything that is necessary to enable reproduction of the results and not holding back essential details as one might well do in business, where commercial secrecy is common and understood. Only meticulous organization in experimental details and tactics (e.g., in verification, confirmation and elimination of alternative causal scenarios) makes this goal attainable.

A very important piece of advice is to organize and discipline yourself as you work,[c] because tracking things down later is a nightmare in most cases.

Perhaps this is true especially for experimentalists, but in any case, it is very important and useful to keep a detailed log-book of everything you do (and do not do!) in the laboratory, even including details that may look silly. Those same details may turn out to be important or even essential later on, and may save you from a lot of wasted time and frustration. Therefore you should take precise notes on everything you do, no matter how tedious this may be.

(We say this knowing that neither of us has fully succeeded in doing this, in spite of aperiodic bursts of enthusiasm, and we both deeply regret this failure. Nonetheless we pass on this precious advice on to you, in the hopes that you will be able to follow it better than we and reap more benefits thereby.[d] Do as we advise, not as we do!)

The bottom line is that being disciplined and well organized is a tremendous advantage for a scientist, so you should make serious and sustained efforts in that direction.

2.8.2 *Be rigorous in your science*

Another important advice is to be rigorous in your work. We recall some of the dictionary definition of "rigorous," namely "rigidly

---

[c]Fair enough, but if any of the readers has ever seen our offices, either live or in a photo, they will have a good laugh and probably not take us seriously any more, at least on the subject of spatial organization.

[d]**Federico:** — Whenever I tried to keep a log-book, I would invariably lose it after some time, or I would become lazy about writing down all the details. I occasionally take notes on separate sheets of paper, and sooner or later I lose those as well (you may feel like laughing at this point, but I can assure you that this is not funny!). I guess this is a pretty hopeless task if you cannot discipline yourself properly.

structured: strict or scrupulously exact: unsparing: severe." The point is
that, in order to avoid unpleasant surprises in the refereeing process or
after publication, you must be more severe with yourself than would be
your severest fair critic.

Challenge your data before others do. If you see an interesting effect
once, try to reproduce it, not only a second time but five times or more. If
you are unsure, ask a colleague to witness your experiment while you
are doing it. If possible, ask your colleague to repeat the experiment in
front of you, to make sure that the phenomenon does not depend on
the observer.[e] Be exhaustive. Do not always look for the simplest
explanation, although often it is the correct one. If there other plausible
explanations, it is up to you to do the tests or calculations to rule them
out, before the referee raises them. In effect, think of the spectre of the
dread referee as a surrogate scientific conscience. (It is true to a
considerable extent that it is the dread of the referee's comments that
spurs people to write papers that are no worse than they are. The threat of
refereeing may well be more effective than the work of the actual
referee.[f])

Be imaginative and creative, to be sure, but make sure that your
procedures are clear and that your data do not depend on instrumental
artifacts. If you find out that they do, be honest about it. Even if you have
avoided the cardinal scientific sin of actually lying about your work, you
would be guilty of scientific misconduct as a sin of omission in not
revealing that some of your published data are doubtful or even possibly
wrong. This is the sort of problem we all stumble on sooner or later, and
there is absolutely nothing to be ashamed of, *unless* you try to cover it
up. Things that you find out later that shake your faith in your published

---

[e]If it does, you can always publish it in the *Journal of Irreproducible Results*.

[f]There is a chess version of this concept (that the threat of attack may be stronger than the
attack if it comes). The bon vivant and well-known cigar smoker Emmanuel Lasker, at
that time World Chess Champion, was to play the nervous master Aron Nimzovitch, who
obtained a concession that Lasker would not smoke during the game. Early in the game
Lasker produced a large cigar and proceeded to go through that pre-smoking caressing
and fondling practiced by many cigar voluptuaries. Nimzovitch appealed frantically to
the tournament director, "Look, look! He promised not to smoke!" "But he is not
smoking, Dr. Nimzovitch." "Idiot! Numbskull! Call yourself a chess player? In chess, the
threat is always stronger than the act!"

work should be published as errata, a reference included with the original publication in your list of publications. (Under the heading of "Ethics in Science" in Sec. 3.4 you will find some more discussion of really serious scientific misconduct.)

2.8.3 Meet your deadlines

While on the topic of meticulous behavior in other aspects, meeting deadlines is another place where being meticulous can open the door to important opportunities. Often perceived as minor nuisance, meeting deadlines is nevertheless important. It is a form of respect for your peers, and often even a necessity. Firm deadlines for submission of applications for grants or scholarships or publications carry their own immediate penalties.

Respecting the deadlines you have willingly assumed yourself will enormously enhance your reputation with your colleagues. Nothing irritates a colleague more than to be promised something which is not delivered on time. Missing too often the deadlines that you yourself have set means that people will stop asking you for the next invited session or grant selection committee membership and your reputation will suffer. To be able to respect your own deadlines, you must also learn (i) not to promise delivery by dates which you will not be able to meet and (ii) to be prepared to drop everything you like to do to meet the deadlines you have accepted.[g]

Scientific deadlines are sometimes more flexible than other ones. However, you should not count on this assumption. We strongly suggest that you always organize yourself so as to meet your deadline, preferably with some time to spare. This can save you from some very unpleasant surprises. In fact, if you have some time to spare, you may be able to

---

[g]Meeting deadlines is important in life in general, not only in pursuing a scientific career. For example, your tax forms should be submitted by the deadline at the very latest, or you will end up paying a lot of interest or even a fine. If you do not pay your electricity, credit card or telephone bill on time, the company may cut your power, and so on. As such, deadlines are a nuisance. However, there is little choice but to take them seriously, as they represent order in situations which would otherwise be chaotic.

intervene if you realize that whatever idea you are submitting is incomplete, or that there is a serious problem with it.

Your supervisor, and later your current employer, will certainly expect you to meet deadlines, which are not necessarily set by them (although some are). These include abstract submission for conferences, submission of papers, reports on your work, and obviously submitting your thesis on time. If you then become a professor, you will have to meet the funding agency's deadline (and probably your University's internal deadline before that), and you will have to enforce deadlines from your students.

In general, your best assumption is to take deadlines very seriously. Aside from avoiding nasty surprises, you should do it simply because it is the right thing to do. It is actually a form of respect towards your colleagues and peers. People who cannot *get their act together* are not very popular in our book, also because they seem to be conveying the message that their time is more important and precious than anybody else's. This type of arrogance in general does not earn you people's friendship, trust or respect. We actually think that flexible deadlines should be largely abolished, because they seem to encourage people to submit late, and to delay the whole process. In the end meeting deadlines, especially important ones, is a matter of fairness and respect for your colleagues.

The issue of meeting deadlines should also guide you in your choice of collaborators. In fact, it would be quite unwise to choose collaborators who are not reliable, and who do not meet deadlines (reliability is perhaps the most important character trait to be considered in the choice of a collaborator, as we argued elsewhere). Conversely, if you are not reliable and are not able to meet deadlines, you are not likely to find scientific peers who are willing to collaborate with you.

2.9 Patience! If you are naturally aggressive, learn when to be patient

As well as brilliance, success in science certainly requires dedication, tenacity and eagerness, but what is less evident is that patience is also required. Training yourself to be patient is certainly not easy, however. It is easy to be patient when the cause of the impatience is not present, but of course patience is most needed when being patient is most difficult. However it is true that, as with many things, patience becomes easier the more you practice and succeed at being patient. There are two kinds of patience to cultivate, the patience with respect to things and the patience with respect to people.

Of the two kinds of patience, patience with respect to things is most essential to an experimentalist (less to a theorist) and is the easiest to learn. In most cases, what is needed is the self-control to step back (at least mentally), take a deep breath or two (or a longer break) and continue at a normal pace and stress level. For example, patience helps a lot when your instruments are not working, because your mind will be able to concentrate better and actually help you in finding the bug (this also applies if you are trying to find the mistake in a computer program). There will be situations in which the delivery of the experimental apparatus you desperately need will be delayed. Being patient in this case will help you find other things to keep you busy and hopefully productive while you wait. This applies as well when navigating all bureaucratic mazes, where a cool head works better than a temper tantrum.

The harder patience to learn is with people we know well, particularly since the situation will probably come up many times. In Italian there is a proverb which says, *"La pazienza è la virtù dei forti"* (Patience is the virtue of the strong).[h] This proverb once again is helpful in everyday life, not just when applied to scientific research. The patience that is required is often simply to hold one's tongue, avoiding a caustic remark and the like. It is best done by seeing the escalation developing and avoiding the oncoming confrontation and slowing or developing the direction of the exchange. In order to do this successfully you will have to cultivate the ability to see the other's point of view to forestall the difficulty. ("Know thy neighbor.")

---

[h]This can be read in two valid ways. One can only be patient with inner strength to control your temper and impatience, or it may be that the strong became so by learning patience. Both are true and worth remembering.

As a student you should cultivate patience with your supervisor, with the research staff and with the secretarial and bureaucratic staff. (As a student you have little power, so you have to be very diplomatic.) It is perhaps a human trait, the desire to have your "superior's" immediate attention, and to be at the very top of their priority list. (It certainly happened to us when we were students.) However, now we can sometimes still feel the same way about our Director. (We often have to fight to get his attention, and the only way to meet him is to schedule a date, time and place with his secretary.) He in turn would like to have more immediate attention from his Scientific Director or even his Director General, or whoever comes next in hierarchy. Therefore it is perfectly normal to feel like that about your thesis advisor and the time that can be devoted to you. Our best advice is, once again, to be patient, and at the same time to insist politely but firmly to get your supervisor's attention until you actually get it.[i]

However, even though your power will increase with success, the need for patience does not decrease much; you have to be patient with different people.

All in all, being patient will help you in getting through the rough times, and believe us, there will be rough times. The only problem is... well, it may be quite difficult to teach yourself to be patient.[j]

2.10   Stand up for yourself!

If you have no difficulty in being patient, co-operative and polite, the odds are that you do not stand up for yourself enough. Even if you do not win all you would like, you will gain respect. As Polonius told his son (in Shakespeare's "Hamlet") *"Beware of entrance to a quarrel, but being in, bear't that the opposed may beware of thee."* But don't waste this effort on trivial causes. (Also as professional athletes know, when you argue

---

[i]**Federico:** — I used to struggle to get my advisor's attention when I was a graduate student, but now that I am supervising students directly for the first time, I realize fully why he was so busy. Managing a group is a lot more complicated and demanding in terms of time, organization and dedication than a student can imagine.

[j]If you succeed in finding a good recipe, we will definitely want to hear about how you did it!

with the referee, it is not to gain this disagreement, it is to have the referee's respect for the next one. In the same way, while you may or may not win the point at hand, you will be better placed for subsequent differences of opinion.)

2.11   Fighting against the odds (You cannot win if you do not play!)

As you probably already realized by now, we like proverbs. A Latin proverb says, *Audaces fortuna iuvat* (fortune favors the bold). Read the following section carefully if you are bold and if you hope to get lucky.

Sometimes you may stumble on an advertisement, and think that getting that certain fellowship or internship or job would be great. Then you look at it more closely and realize that perhaps the competition is going to be extremely tough, and that your chances of getting the position are quite slim. Similarly, you may want to apply for a certain grant, but then when you look at the statistics you realize that the chances of getting it are very remote. Many people we know, in these cases, will simply give up without trying. There are a lot of people out there, who cannot bear the idea of failure, and who are too scared even to try. Or there are some that may say (if there is a chance of losing) "E'en that would be some stooping, and I choose never to stoop."[k]

Your estimate for the competition may be quite wrong. Even if you know that the ideal candidate for a certain competition is going to apply, making your chances of success extremely low, that person may have other offers and decide to give up on this particular competition, or that ideal candidate may have an accident on the day of the interview, or may get the offer and then turn it down unexpectedly. In each of these cases, if you did not apply because you were sure that this person was going to get the job, you will feel stupid afterwards. If this happens, you will have killed by yourself one possible chance of success. Since scientific research is an extremely competitive field, the chances of success in general are low enough that you are better off not shooting yourself in the foot just because of your pessimism or lack of courage.

---

[k]From Browning's grim little dramatic monologue poem "My Last Duchess."

If you ever find yourself in that situation, we strongly advise you to *fight against the odds*, and give it a shot anyway. As the Scottish golf professional will say, of a too-tentative putt, "Never up, never in." In other words, your odds of getting a job go up if you file an application! In the worst case, you will not get an offer, which leaves you in your original situation, no worse. Perhaps you will have spent some time in preparing the application or the grant proposal, but you should think of that as time that you *invested* in yourself, not wasted. There will come a time when you will be able to use, or recycle, that particular attempt. Yes, you will have failed for that particular case. Perhaps your ego will even feel a little bruised. Set it aside: it does not matter, because everybody, even the greatest minds will fail every now and then. In the best case on the other hand, you will get the offer for which you were hoping, in which case you can choose whether to follow through or not. In other words, you will have created a new opportunity for yourself.

Let us give you two specific examples, which hopefully will convince you that what we are claiming is worth listening to.

Federico: — There is an exchange program between Europe and Japan, called REES (Research Exchange for European Students). Only students from France, Germany, Italy and the United Kingdom are eligible to participate in this program (other countries may have different programs, of which we are presently not aware). Each country sets its own rules for the competition. On average, each country sends 3-5 students to Japan during the summer, from the beginning of July to mid September, to do a research internship in a Japanese University or government laboratory or private company. This is the program that landed me in Japan in the summers of 1999 and 2000.

But let me start from the beginning. The first time I heard about this program was at the beginning of 1998. At the time I had just finished military service, and I was preparing for Ph.D. admissions exams in various Italian universities. I applied for a fellowship within this exchange program, thinking that it would be interesting even just for the experience of spending a few months in Japan. There were four positions, and I heard later that more than thirty candidates applied. I was close to getting a fellowship, but did not make it.

Then I started my Ph.D. in Rome, and when I saw the ad again the following year, I decided to give it a second try. This second time I had more time to think about the project I wanted to work on, and I was also able to find a laboratory that accepted me (provided that I got the fellowship). This particular instance proved very useful, because it gave me bonus points in the competition. Again there were four positions and roughly thirty applicants, and this time I ranked first.

When I came back from Japan in September 1999, although my experience had been quite stressful in many ways, I decided that perhaps it would be good to go there a second time. When I submitted my report to the Institute that was funding the program, I asked if it would be possible to apply one more time. The person in charge answered me politely that in principle there was no rule to forbid that. However, he added, it obviously made more sense to send *new* people, which implied that if they had again four fellowships and, say, five candidates including myself, I would be the one left out. I decided to apply anyway. After all, it was not going to take up much of my time. I just had to update my CV (which was progressing slowly at that time anyway) and to come up with another viable project. Once again I got the acceptance letter from the same lab where I had worked during that summer, although the group leader was extremely skeptical that I could get the fellowship a second time.

Well you already know that I did go a second time, and by now you have probably guessed that I won the fellowship again. That particular year, the Institute was not able to advertise the program properly. In previous editions the program attracted thirty candidates on average, whereas that year they received only two applications, including mine! There were still four fellowships, so I got the chance to go a second time, to the surprise of all the people in Japan (who probably thought I had cheated in some way).

Another good example is the ad that ultimately brought me to my current position. During my work in Denmark, at some point while reading *Physics Today* and other similar journals, I noticed that the number of job offers had increased enormously with respect to previous years. Although I had recently finished my Ph.D. and had a very thin publication list, I decided to check out the job market and apply in many places, particularly in the United States. Even though I did not expect to succeed, when all the rejection letters started arriving I was of course somewhat disappointed. Since I had received so many offers for post-doctoral positions, I thought that maybe I could get into a faculty position somewhere, thereby accelerating my career. Looking back, my CV and publication list were definitely too weak to give me a fair chance in a job market that was still extremely competitive, in spite of the sudden surge in job offers. In retrospect this was a good thing, because I certainly had neither the maturity nor the experience to build my own group and set up a laboratory.

Then I saw the advertisement from *Institut National de la Recherche Scientifique* (which I had never heard of), located in Varennes (also completely unknown to me) near Montreal in Quebec (these sounded more familiar), the francophone province of Canada. The ad was particularly interesting for me because my girlfriend was already living in Montreal, so this job offer could have given me the possibility to join her there. I discussed it on the phone with her however, and I did not hide my skepticism about applying, since I had been systematically turned down when applying in the United States. I still remember her words, *"Don't bother,"* which suddenly rang a bell in my head. *Wait a minute*, I thought, *didn't I fight against the odds and win when I applied the third time to go to Japan?* I decided to apply anyway, even though I thought my chances were slim, at best. It was April 2001, and by the end of May I received an email asking me to cross the ocean for an interview. During the interview, I discovered to my great surprise that there had been only ten applications for that position, and that *I was the only one summoned for an interview*. Once again, I had fought against the odds, and eventually I won.

I urge you to do the same, because as you can infer from my personal experience, it can make a huge difference in your life. Now I live with my girlfriend in Montreal, and if I had not gotten a job here, it would have been extremely difficult to continue our relationship. Not only that, I made a major step forward in my career, moving from a post-doctoral job to a faculty position. I negotiated a late start so that I could get some more post-doctoral experience and publish some of the work I was doing. This gave me enough experience and maturity to start off successfully as a young faculty member.

2.12 Nothing succeeds like success

Believe in yourself, within reason, and you will be more successful, not only in science but in life in general. How can you expect someone (such as a future employer) to believe in you if you are diffident about yourself?

Since scientific research is an activity in which resources are extremely scarce, in most countries where it is taken seriously, it is governed by a strongly *capitalist* system. As one of our senior colleagues would say (in a remark usually made about banks), *"On ne prête qu'aux riches."* (*They only lend money to rich people* (with the implied clause, *"who don't really need it"*.).) This means that if you are successful in getting a grant, when you apply for the next grant, your chances of success are higher. Nobody on a grant selection committee will say, *"This person already has a lot of money, let's give someone else a chance."* Rather, they will say, *"This person has a strong track record. Other funding agencies are investing in this scientist. The applicant has brought to completion several important projects. We too should invest in this person."*

The concept *"nothing succeeds like success,"* incidentally, applies to most fields of human endeavor, not only to scientific research. Part of this is due to psychological reasons: the more you succeed, the more you learn to believe in yourself, perhaps to become bolder, and this undoubtedly helps you in your future enterprises. Therefore another important piece of advice is that if you believe in yourself, your chances of succeeding are much greater.

There is, of course, a *caveat* here. Believing too much in you may not be realistic, and may even turn out to be counterproductive. This in fact can be in conflict with an earlier piece of advice which said, *know yourself.* A simple example is that if you are 35 years old and have never run a marathon, even if you are in relatively good physical shape, it is unlikely that you will become the *world champion* in marathon competitions.

The bottom line is, it often helps to be optimistic, and it is good to be realistic. You are certainly better off *not* being pessimistic.

2.13   Europe vs. North America

Europe tends to be more hierarchically structured than North America, and particularly so in University environments. In countries like Germany and Italy, for example, Full Professors still wield enormous power, and can essentially decide on who is going to be hired or promoted in their departments. Senior professors seem to enjoy a special status and exceptional powers almost everywhere in the old continent, from Spain to the UK, from France to the Netherlands to Denmark. Of course every now and then you will meet a senior European professor who is "enlightened," and who gives a lot of freedom to his younger associates. This, however, tends to be the exception. If you are a young dynamic scientist, wishing to start your own research activity, this lack of opportunity to choose your own path in your early career is a very good reason to leave and look for better opportunities elsewhere. The European research system urgently needs serious and radical reform in this regard, both at the level of individual countries and at the level of the Union.

The European mentality is really quite different from the North American one, and this is naturally reflected in research policy and in the ways universities and departments are organized in the two continents. (As usual, one of the key issues is money, although it would be too simplistic to describe this gap as a purely financial issue.)

In Europe it is not uncommon to land a permanent position at a relatively early stage in your career. This is, to some extent, an advantage. The disadvantage of this system however is that a junior faculty member or a junior staff member in a government laboratory will usually be part of a group, with a senior member acting as a boss. This means that you are essentially trading scientific independence for job security.

In North America the situation is essentially the opposite. You will normally have to fight hard to obtain "tenure" in an American university, but you will be running your own show from day one, i.e. you will be completely independent. This means that, as long as you are able to secure the funding necessary to carry out your research, you will not have to comply with a senior scientist's decisions. Rather, you will choose your research topics and take full responsibility for your choices. In Canada the tenure struggle is said to be less severe, but the independence in research is real as in the United States.

As was discussed previously, another important difference is that North America tends to be a lot more open to hiring foreigners for research, both for short-term contracts and for more senior and permanent positions. Perhaps because North America (essentially Canada and the United States), is intrinsically a land of immigrants, it is generally more open than Europe to hiring foreigners. In Europe it is perhaps common to hire scientists from other European countries, but it is very rare to offer forms of long-term employment to people who are not citizens of European countries. This, together with the high levels of funding (especially in the U.S.), has given a great edge to the North American system throughout the last half century. The trend started before the second world war, when many famous scientists (such as Einstein and Fermi) were fleeing Europe (many because of anti-Semitic laws in Germany and Italy, and because of the imminent war). Presumably the success thus obtained has maintained the inclination to be open to foreigners even when anti-Semitism is no longer a factor. (However more recently, under the impact of the "Fortress America" Homeland Security mentality, American immigration has become much more rigid and a significant deterrent in attracting scientists to the U.S.)

Each system thus presents advantages and disadvantages, and you should decide which system you prefer based on your personality and ability and on your chances of landing a good position.

If you want to be a leader from the early stages of your career, say as an alpha scientist, as we have termed it, then clearly starting your independent scientific career in North America is the right choice (providing you succeed!).

If, on the other hand, you prefer having job security early on, perhaps because you are a natural beta scientist (or suspect that you are a gamma scientist) or perhaps because you want to start a family and do not want to live with the fear of being unemployed in a few years' time, you are better off seeking employment in Europe (especially if you are European). The European system offers much more in terms of permanent positions for beta scientists.

It may perhaps not surprise the reader to find that, as scientists working in Quebec in Canada, we find that Canada (Quebec in particular) is a good compromise between Europe and the United States.

(Incidentally, many non-scientists find Canada to be a good compromise in terms of the overall quality of life, and the people's mentality.)

You have the possibility to run your own show from day one, and at the same time your chances of being tenured after a few years of hard work are quite high (as mentioned briefly above, many people say that *everybody* gets tenured in Canada, and, although this is undoubtedly an exaggeration, it gives you an idea of what the situation is).

Having discussed the advantages and disadvantages for the young scientist of Europe versus North America, an interesting strategy is available for a European scientist of alpha caliber. That strategy is to develop your research career in North America to a high level and then return in triumph to Europe to one of those plush full professor or research director positions. (Of course, if the competition for the American prize proves too difficult, you may have also missed out completely on the junior "starter" positions in Europe. Nor is there any guarantee of a plum position in Europe unless you have become a really internationally prominent figure in your field.)

2.14   Working in Asia (Federico's experience)

Federico: — Japan has been for several decades the second largest economy in the world. Having invested continually for many years in innovation, science and technology, this country has been a leader in some fields of research for several decades. However, the Asian landscape in science and technology has been changing markedly. Countries and City States like South Korea, Singapore, Hong Kong and Taiwan have also consistently invested in research and development, and this effort, begun well after the Japanese, is now bearing fruit.

More recently, China and India (the two largest countries in terms of population) have also begun to invest significant resources in scientific research, because of their desire to join the ranks of the world's most developed and industrialized countries. These big players are likely to increase their investment in numbers in the near future, and they have a lot to invest. They also have the advantage that manpower is fairly cheap. The situation is such that a reverse brain drain of sorts has already started between North America and China, mainly of Chinese who were in the pipeline, so to speak, in North America. Also many Chinese who had succeeded in science in the U.S. were recently attracted back to China with large sums of money (both their personal salaries and research funding).

Certain regions in Asia, while still developing from an economic point of view, are populated by friendly people and benefit from a warm climate all year round. This is partly what motivated J., a scientist from Southern Asia, to move from Europe to South East Asia. Other reasons for his move include the proximity of his family, now only two hours away by plane, as well as the "comforts" that his wife and child can enjoy on a campus that is fairly modern and safe. The main drawback he is presently facing is that now J. has limited access to funding and infrastructure. Since he is smart and resourceful however, he is not letting this situation deter him. He decided to base his research program on sound, simple ideas which require limited funding to be carried out. At the same time, he uses his previous contacts and collaborations to "outsource" whatever use of infrastructure he may need. Since we live in an environment (actually, a whole continent!) which values money above everything else (not only in research), we find it refreshing to still find scientists like J.

European and North American scientists have in the past considered working in Japan at some stage in their careers, and more rarely in e.g. South Korea or Singapore. Soon, however, the temptation to work in China may prove to be too strong to resist. (Apart from the work experience itself, this could yield ancillary benefits in the form of good contacts for obtaining Chinese post-docs!)

In Asia, scientific communities tend to be much more hierarchical than in the western world (just like most Asian societies). I found this to be particularly true in Japan, a country where I worked for two consecutive summers and which I visit fairly regularly. However I know very few foreigners who relocated permanently there. This is not to discourage you if you are thinking of working there, but just to warn you that the cultural differences between Europe and North America are trivial when compared to those between the Western World and Japan. (This probably applies to most of Asia, for that matter, although I do not have a direct work experience with other Asian countries and cannot give a truly informed opinion.)

I still remember vividly the first time I landed in Japan (it was also my first time in Asia). While getting off the airplane, my first thought was, "Oh my goodness, this is a different planet!" In the end I liked it very much, so much that I went back there again as a summer student a year later. Now I am happy to visit there whenever the opportunity arises, and I still have many friends there. Nonetheless, I do not think I could live in Japan over the long term. It is just too different from anything I know and am used to. It would be next to impossible for me to integrate in such a different society.

The hierarchical system holds for the industrial lab where I worked, and, of course, for government labs in general. (Of course these last tend to be hierarchical everywhere, since they are "managed" environments, in the sense that managers will tend to set your priorities for you and define your projects, unless you can convince them otherwise.) Alas it is also painfully true for universities as well.

A few years back, a friend of mine, S., went to Japan to spend a sabbatical year. The idea was to collaborate with a local Full Professor. Being an Associate Professor himself, S. was shocked to discover that he always had to go through an intermediary to communicate with his host, because of his perceived lower "rank." This proved to be very inefficient, as well as frustrating. This working relationship certainly did not sound like a real collaboration.

Of course, hierarchy has advantages and disadvantages. Just as for most things, it is a strictly personal decision, whether you would fit into that type of working environment or not. (End of Federico's experience.)

2.15 Brain drain vs. brain gain

Many countries that will not or cannot invest significant resources in scientific research naturally suffer from what is called the "brain drain", as their talented youth leave to employ their talents elsewhere. (Indeed many of our readers are likely "brains" that were drained from their home country, and flowed into North America or Europe — and so indeed is one of us.) In planning your scientific career, you should take this aspect into account. If you are a successful scientist, you will have many job offers, but the truly interesting ones that represent good opportunities will be limited to some very specific geographical areas, the attractors of the "brains."

2.15.1 *Two brain gain countries: United States and Canada*

Most of the brain drain so far has occurred towards the United States, where the budgets for research are still immense, at least when compared to any other single country. (Naturally the competition for these enormous funds is also particularly fierce there.) Since the brain drain is a natural by-product of the modern capitalist society, for faculty positions the U.S. departments are able to offer more money than others in terms of salaries, start-up funds, and so on. This enables them to attract the best scientists in the world, in every important field of research. This is, to some extent, unfortunate (from the point of view of other countries), but it is reality.

While a brain-gain country with respect to the rest of the world, Canada itself has suffered significantly in the past from the brain drain from (or through) Canada towards the U.S. Canada is thus trying to reverse this trend as part of its *Innovation Strategy.*[1] In this sense, it is making an enormous effort in terms of brain *gain*, and is managing to

---

[1]Canada's innovation strategy includes for example the Canada Research Chairs program, which is specifically designed to attract world leaders (senior chairs) or potential world leaders (junior chairs) to Canadian Universities. More information on this program can be found at: http://www.chairs.gc.ca/english/About/index.html. Other countries have similar programs. Examples include EPSRC and Royal Society Fellowships in the UK, SFI Fellows in Ireland and Federation Fellows in Australia.

attract many bright foreign scientists, besides luring back Canadians that had gone abroad. This effort is not to be taken lightly, considering Canada's relatively small population of only 32 million people. The Canadian strategy is particularly effective in our opinion, because it is also relatively simple for a foreigner to acquire permanent residency here first and then citizenship, and to integrate in this country, which is precisely a land of immigrants. It is easier to integrate here nowadays rather than in the U.S.

Incidentally, in the aftermath of recent terrorist attacks against the United States (in particular those perpetrated on Sept. 11th 2001), the U.S. has become increasingly more selective in admitting foreign students and scientists into the country (perhaps rightly so, although in the long run this attitude will become very damaging), and this has had the net effect of increasing tremendously the number of applicants (students, post-docs and professors) to Canadian universities, industries and government laboratories.

2.15.2   A brain drain country: Italy

Federico: — Italy, my home country, is one which is suffering tremendously from the brain drain. The working conditions there are simply unbearable for a young scientist who wants to pursue an independent scientific career. Italy invests less than 1% of its GDP in research. If we compare it to France, for example, this is a ridiculous budget. France spends more than 2% of its GDP on research, and since its GDP is about double that of Italy (whereas the population is roughly the same), there is a total difference in budget of a factor of four approximately. Indeed one finds Italian scientists all over France. It is common to find them also in other countries that invest even more in scientific research and leadership, like Canada, Germany, the UK, and the United States. In view of all this it is truly surprising how often one sees Italians in Italian research centers managing to publish in first-rate journals (albeit particularly in the fields in which the running costs are less, such as theory and modeling).

In recent years a group of young Italian scientists has interviewed a great number of Italians working abroad, and has published a small book called *"Fuga dei cervelli"* (brain drain). This book contains all the most basic reasons that may induce a young scientist to leave his/her home country to pursue a career abroad. Indeed I could identify very well with many of the stories reported in that book.

I think that, in the long run, Italy's lack of vision and its inability to initiate an Innovation Strategy like the Canadian one will simply lead to a continuing slide in Italian front-line research. I suppose however that my home country still has to go through the painful process of becoming a real democracy, before it can focus on more long term problems like fostering basic research. In essence this means that as long as I am a successful scientist, and my profession is important to me, I am not likely to go back to Italy at a later stage in my career. Incidentally, I have accepted this state of things a long time ago, and I was well aware of this when I first left my home country in the Fall of the year 2000. Although these reflections are very personal, they may well apply to you too. When you leave your home country looking for better opportunities abroad, you should be aware that if you are successful, you may well never want to go back, particularly if the gap is increasing with time. (End of Federico's discussion.)

2.16 Knowing an extra language may help enormously

Because we believe that you should write your thesis in English (see Section 5.3), you may think that we would advise against learning other languages. This is not true. If you have the time and the interest to learn an extra language, this may become an extremely useful skill not only in your profession but in your life in general.

Fair enough, English tends to dominate in various fields of human endeavor, not just scientific research. Yet you may find yourself in situations in which knowing another language may be important or may even save your life.

Federico: — I have an anecdote about this, which perhaps has nothing to do with science, but which I find useful and interesting. When my father was a teenager, an American boy of Italian origin came to spend one year with his family, on an exchange program. During this period, he learned Italian. This was in the late fifties. Several years later, he was drafted in the U.S. military, during the Vietnam War. However, since he spoke Italian, he was sent to a base in Italy, instead of going to Vietnam. This may well have saved his life. In any case, it saved him a lot of trouble.

Incidentally, I probably got my present position because the Institute here at Varennes in the province of Québec was looking for an expert in Nanoscience who could also speak French, and this narrowed down their list of applicants tremendously. So I am quite convinced that my knowledge of French (far from perfect, mind you) greatly helped me to get my current job. (End of Federico's anecdote.)

Unless you intend to live in China however, we are not yet suggesting that you learn Chinese (although it may turn out to be quite useful in the decades to come). We would suggest rather investing some time in learning other European languages, which are reasonably easy to learn if your mother tongue is also a European language, or if you already speak fluently another European language. (You might well choose the language of a country you enjoy visiting.)

Tudor: — A friend of mine learned a smattering of Italian to profit more fully from a sabbatical leave there, kept up with his study on his return, and, now that he has retired, visits Italy almost every year.

2.17   Keep yourself up to date

Once you choose a research program, you will be tempted to continue working on it for a long period of time, particularly if you are successful. (We actually happen to know some scientists who have been working in the exact same field for 2-3 decades, if not more; they are undoubtedly the world's foremost experts in their field. To be quite honest however, we feel somewhat sorry for their rather narrow niche existence.) It seems inevitable now that the lifetime of a specialty will be less than a typical lifetime in science, so that this mono-mode existence will become less and less feasible. In all probability the scientists now training will be recycling themselves two or three times at least.

This is, of course, good and bad at the same time. It is good, because it will give you a chance to delve deeply into this topic, to make it yours, and to become a world expert on it. If you reach this level, acquiring new funds to continue this line of work will become increasingly easy for you, since you will be considered one of the foremost experts in the field.

At the same time, if it keeps you from starting new projects, and from broadening somewhat the scope of your research, it will also be bad. It may become somewhat difficult to continue being creative while working on the same subject. It becomes repetitive and it quickly loses its initial fascination, which was perhaps the main source of drive that should keep you going and keep your enthusiasm at high levels.

There are priorities and fashions in funding so certain topics that lose their priority can be gradually cut and even come to an abrupt stop. Therefore, being able to change subject is not only important to keep up your enthusiasm and creativity, but it may actually become necessary to draw the funding necessary for carrying out your research. Thus, in doing research, being versatile and flexible will be an invaluable asset, especially in the long run. Be open to new ideas, and to learning new skills. Research is about making new discoveries, facing new challenges, not re-cooking the same stuff over and over again.

It is very useful to spend some time reading about the general trends in scientific research, funding, and about your discipline in particular. Keeping abreast of funding trends in your own field of research is also important.

In today's modern world, gathering information is of paramount importance. However, it is also important to filter it properly, and to discern useful information from the rest.

Chapter 3

The Game of Science

Sections of this Chapter

3.1 The ecology of science
3.2 The peer review system
3.3 Ethics in science
3.4 When ethics fail
3.5 Intellectual property rights and patents
3.6 Gender-equal opportunity employment

3.1 The ecology of science

Most young scientists just see the part of the science environment or "ecology" that has become evident to them in their contacts up to the present and the various roles which they have played. These various possible roles include the following: student, post-doctoral fellow, submitting author, referee, grant applicant, grant selection committee member, employment selection committee member, journal associate editor (or even editor), workshop or conference organizer and many more. The overview provided here is normally only obtained after many years of experience in the various roles a scientist may play. Most of the remarks are limited to public science, with only a few remarks addressed to industrial science and to government science. From this overview it will be evident that you need to be very aware of how the system works in practice.

Many who have succeeded in science may find these sections "cynical," and wonder whether the illusions of beginning scientists should be so challenged. The adjective preferred here is "realistic,"[a] and we believe that this realism is essential if beginning scientists are to avoid the many pitfalls that beset their early career choices. The need is to know the world of science as it is, not as how it "ought" to be. The common platitudes about the rigor and even-handedness of the system will not do here.

In the ecology of public science, the equivalent to the necessary biotic energy of natural ecology is (of course) money (or "funding" as it is usually and more delicately called). The equivalent to the solar energy input is the flow of public funds partly through the funding of university positions and students (bursaries, scholarships or the like) and partly through the direct funding of research projects of many kinds. As in natural ecological systems the object of this multiple-player game for each player seems to be to maximize the amount of funding one attracts to oneself. Essentially there is an overall Darwinian ethic of "survival and growth of the fittest and the most eager."

While most of the competing scientists ('players" in the "game of science") tend to say, if asked, that all that is needed is "enough to do good work," in practice it seems that the appetite for funding is effectively insatiable. However, unlike the players in a pure Darwinian system, some few are indeed happy with less than what they might get. One might guess that this is partly because this under-funding provides a rationale for not pushing their research to the utmost, and partly because they do not want to engage in the quasi-political campaigns needed to obtain the largest grants.

Now the government can and does control the general areas of funding by introducing many *ad hoc* programs and funding them according to government policies and priorities. (Currently "hot" subjects such as, for instance, nanotechnology, get more funding, while energy research tends to correlate nonlinearly with the price of oil.) The science "players" can choose or not to go into these areas, which choices can be regarded as strategic decisions. Within each area, however, the

---

[a]People referred to as "cynics" nearly always say that they are "realists". By inference then, the critics of the "cynics" have "unreal" expectations and are "naïve" or perhaps are "idealists." We are no exception to that common position.

funding is generally to be awarded on the basis of "excellence." (Like pornography, excellence is hard to define, but everyone says that they know it when they see it.) Many other decisions are also made on the basis of "excellence": fellowships, scholarships and publication in refereed journals.

Putting aside for the present the philosophical question "what is excellence?", the immediate question is "How is "excellence" (whatever it is) to be judged?" Of course the naturally most qualified judges would seem to be the appropriate subset of the experts in the field ... But aren't these the people who are going to get the money? (For those who, like us, enjoy Latin proverbs, this one applies: "*Quis custodiet ipsos custodies?*" i.e., "who guards the guards themselves?") Is this not a basic conflict of interest? Well yes, this certainly poses a problem.

This conflict of interest is here solved empirically by time-sharing, with the scientists playing at different time the role of claimant and jury. (The sentencing judge with the executive power is the funding agency or journal editor to whom the "gatekeepers" report.) In effect, a scientist being judged by a jury of scientists this week will be on a jury judging other scientists next week. This request/evaluation process is known in English law as "trial by a jury of your peers," or according to a satirical history of England called "1066 and All That" (W.A. Sellars, R.J. Yeatman, Methuen (London) orig. 1930, repr. 1953 p.26), in discussing Magna Carta, "Barons should not be tried except by other Barons, who will understand." In science this evaluation process is called the peer review system or simply "peer review," a process which takes many forms, depending on what is being judged. Peer review is sufficiently important that it is discussed in considerably more detail in the next section.

Now, in a curious example of circular reasoning, it transpires that "excellence is defined implicitly as that quality which is evident in the work chosen by the ensemble of peer review structures created to choose excellence."

In this curious and competitive contest or game, with ill-defined rules, where the players judge each other, the player scientist who wishes to succeed (at best) survive (at least) at a comfortable level should understand how the game works.

The fundamental aspect is the *publication game* where the gains or credits are accumulated in the form of *refereed publications* in the player's CV. (Anyone reading this work is unlikely to require to be told that "CV" stands for the Latin *"curriculum vitae,"* the "life work" of the individual.) The publications are only listed in the CV's but in fact they are implicitly weighted by the reader or evaluator in ill-defined ways by the impact or importance of each publication, by the perceived importance of the work (which may take years or even decades to emerge) and by the *impact factor* of the journals in which the publication appears. The CV is thus a key element in ill-defined contests for scholarships, fellowships, promotion and research grants and the like. Other quasi-publications flowing from this are in the form of un-refereed publications (from conferences and the like) of invited conference talks and reviews (especially if refereed, and even more so if published in well-known journals where they become refereed publications).

The *publication game* which leads to the creation of the CV is thus the basis on which all this progress for the scientist rests. It is therefore essential that the scientist learns to play the publication game to maximize the opportunities which successful science should afford. All this is conventionally and sardonically summed up in the phrase, *"Publish or perish!"* While "publish" here essentially means "publish original work in the best refereed journals" it is not limited to that aspect. Reviews, workshop proceedings, books and oral presentations that are then published, all, refereed or not, are a form of publication and thus of the building of a scientist's reputation.

Do not forget, however, that more ephemeral presentations given as part of a selection process or indeed seminars in general (not to mention conference presentations, including posters) can also be very important in particular cases not related to the building of the CV. Although ephemeral in the publication game and the building of your CV, these events can be invaluable as means of getting yourself more directly known to possible future employers, to selection committee recruiters, potential referees of your work or grant applications and even to post-doctoral candidates and students that you might want to recruit.

The beginning scientist is a bit like a beginning actor. You have to get cast in a good part (i.e., a good research topic), you have to do it well (and be perceived to be doing so by colleagues as well). Finally the production has to be a hit as judged by the public.

For a scientist the publications are the productions for an actor, and like, say, films, one can go for a limited release (equivalent to a specialized journal not read much by the non-specialist) or go all the way to full-blown world release (equivalent to publication in *Nature* or in *Science*), with a full range in between.

The style and shape of the publication is determined in effect by the choice of journal.

In an ideal world, providing a paper contains well written solid work, where it is published should not matter, since only the science should count. In the real world if your stuff is really good, you want it published in the journal where it will have the most impact. It is also true that the average effect (measured if necessary by a well-defined *impact factor*) of the journal has a large impact on the perception of the work published. People (including those making important decisions on your future) pay far more attention to a paper in Nature, and will assume it must be important, according you the benefit of the doubt on things that seem obscure to them.

Given this climate it is natural therefore that you want to optimize your effort by publishing in the best journal that you can. Aim too high and the risk of rejection is high, aim too low and, while acceptance may be easy, you are, so to speak, giving away "impact points." Part of the game of science publishing from the author's point of view is in being able to estimate which this optimum journal might be.

There is a growing tendency to quantify scientists' and institutions' worth by calculating their citations and the impact factors of the journals they publish in. Although the aim is to drive scientists towards a higher quality, this is a dangerous game, since the indicators being used are not always objective and reliable. Citation indices (which indicate how often a paper has been cited) are not useless (after all if a paper is rarely cited it is hard to see how its impact can be significant). A paper can, however, be highly cited simply because it makes a significant correction in a standard value of coefficient, while a really original paper, with considerable initial impact judged by the work it stimulates, may be superseded by a paper which extends the work a bit (to be sure) but also presents the concept and its significance in a way that makes the work somewhat easier to grasp. (A really original paper is often difficult to read because deep originality does not always go hand in hand with the crystal clarity needed to make the deep ideas plain.)

3.2 The peer review system

Having outlined the ecological system of public and published science, it is time to discuss the key component — the peer review system. Depending on your personal experience and views, the following statements may either sound obvious or shocking to you. One point to retain in discussing the peer review process is the same as that for the justice system, namely, "Justice must not only be done, but be seen to be done." (A cynic might add, "Therefore, if injustice is to be done, it must be seen to be just.") This is the main point against secret trials and closed-door justice. Yet the essence of peer review is that the deliberations are kept secret, and in many cases, the identity of the judges as well. How then is a submission for publication to be judged fairly? That is the basic dilemma of the peer review system as we have it today. It seems likely that the peer review system of today is (like democracy) highly imperfect, but all other systems attempted so far are disastrously worse. We will now go into this in some detail.

It is clear that there is an inherent conflict of interest in the peer review system. Look at it first from your point of view when you are presenting something (a submission for publication or even your whole career) to be judged. If your work (papers and grant proposals) is to be judged by peers in your same field of research, it is very likely that one or more of your reviewers is working on something very similar to what you are doing. Often as not they are working on essentially the same topic, trying to get their paper out first, and perhaps competing for the same pot of money. Being human, it is not easy for that person to give you a fair review, even though they feel that of course they can put their bias aside. On that basis, it seems that these competitors ought to be excluded from judging your work.

Now look at it from the general science point of view of getting the best reviews of your work. If everybody who might be a competitor is excluded, how can an expert (as opposed to uninformed) opinion be obtained? The basic dilemma is where and how to draw the line between sufficient expertise and conflict of interest.

By the way, do not forget that you yourself will also be in that position, i.e., of judging a close competitor, at another time. Ask yourself these questions. How fair can you be? Should you recuse yourself (as judges are required to do) for explicit conflicts of interest? Be very scrupulous in such cases, if only to become known for being scrupulous.

In this context, trying to avoid real conflicts of interest without excluding all the experts, many journals and funding agencies will ask you to submit with your manuscript a list of possible reviewers for your manuscript or grant proposal. (There will be some general guidelines so as to avoid blatant conflicts of interest in preparing your list. For example, a frequent collaborator, your previous thesis advisor, or supervisors, or people from your lab etc. should also be excluded, and so on). This will still give you the possibility to come up with a list of scientists who are your friends, and who are therefore likely to give you a *fair* review.

Speaking of fair reviews, the classic picture for Peer Review is another cartoon by Sidney Harris,[1] shown here with caption, *"That's it? That's peer review?"* (Permission received for the use of this cartoon from his web site ScienceCartoonsPlus.com.) Of course, this would have been easy to describe in words, but the visual impact of that great crossing-out of this large equation is too good not to show.

Most scientific fields are small, and therefore the immediate community you work in will be essentially divided between people who are your friends, those who are your enemies and those who are genuinely disinterested. As one senior colleague once said, the reviewers in your list should be experts in the field, and they should be friends. If you have doubts whether someone is a friend or not, or you think they may be struggling to get to a certain result before you, you are better off *not* suggesting them as a referee. This is all very well, but what you should also realize is that these referees will be considered by the committee as "your" referees and thus anything positive will be somewhat discounted, while anything negative from them can be very damaging, much more so than for a referee you did not choose. Be very sure you know who your supporters are. This is certainly a case of "Better safe than sorry." More importantly, "Know thy neighbor."

Note that while we specified what you might think of as a *fair* review, that does not necessarily mean a review that serves science well; it might well be not very rigorous. You can expect that your friends will not try to shoot you down intentionally, and they may be a lot more lenient than others if you did not make serious mistakes.

In this light, consider the strategy of D., a colleague from another university than ours. While D. was writing his grant proposal for one of the main funding agencies in his country, he once met a senior colleague to hear his opinion about the draft proposal. The senior colleague liked the draft very much, and proceeded to point out several minor inconsistencies that he thought should be corrected. In the section on "Collaborations" for example, he asked why D. had not written down the name of a certain professor, whom he knew to be a friend of D. and who was likely to be a potential collaborator. D. replied that, since this possible collaboration was still far off in the future, he would rather list him as a possible reviewer, rather than a collaborator, bringing him an immediate benefit. This sharp (and perhaps cynically tactical) reasoning greatly impressed the senior colleague as being unusually perceptive for one so young.

Most journals and funding agencies will allow you to ask that particular scientists (with whom you are clearly competing, and perhaps with whom you have had a recent row) be excluded from reviewing your contribution. (You may find that, although this policy exists, you may not be told that it is in place (and will have to ask), and will not be prompted for a list of reviewers to exclude. This information is usually in the fine print of the detailed rules.) If you are aware of such dangerous people, do not hesitate to provide such a list, but use this defensive tactic with discretion and do not make the list too long. (Overuse will tend to lose you credibility; the reputation of being considered an isolated "crank" is hard to live down).

It sometimes happens that, after being excluded from a committee or having an opinion overruled or discarded by a journal editor, a person may overstep the unwritten rules and write nasty comments to the journal editor or to the grant selection committee. Not being in the review process (and thus not covered by anonymity of the review process), you may get to hear of this. If you do (usually unofficially) together with the name of the person, be happy, because it identifies such a person so that you can take steps to exclude them explicitly in the future. (Do not complain to the person in question, since that is guaranteed to be counter-productive. Naturally you should not react to the review committee, since you are not supposed to know what happened.) Since any excesses inside the process itself are supposed to be covered by the anonymity of the details of the review process, you will rarely get to hear of them, and that only unofficially.

To learn more on the merits and pitfalls of peer review (and on many other interesting topics), journals like Science and Nature frequently discuss them in their News Features, Opinion Articles and Correspondence Letters.

Various forms of anonymous peer review apply for submissions in different contexts. In order of weight and consequences they range from the relatively frequent application to the peer review process for (A) submissions to refereed conferences, and (B) submissions for refereed journals, to the rarer events comprising (C) applications for research grants from various sources, and (D) applications for scholarships, fellowships awards and the like. The processes for publishing submissions to journals and refereed conference proceedings (A and B) and for submissions to selection committees (C and D) are rather different and will be treated separately. The simplest will be discussed first and that is the jury-like system for C and D.

3.2.1 Peer review in open committees with unpublished proceedings

Some of the most important peer review is partly open. In those situations (C and D above), as is usual in other types of juries, the membership of selection committees for grants, fellowships, employment, prizes and the like are known in advance. However the details of deliberations are not to be divulged; only the final decisions are to be made public or communicated to the applicant, as the case may be.

With known members in a committee setting (with explicit exclusions for well-defined conflicts of interest), there is some safety in numbers, since a single extremist will in effect be moderated by the consensus. Also in a committee an extremist in a particular case will usually (consciously or unconsciously) tone these opinions down as to maintain credibility with respect to the rest of the committee and for other candidates. (By the way, this is the most important reason why documents to be looked at by a committee should be written in a particular way. The text should always be written both to convince (or at least disarm) the expert and to prove appealing to the moderately well-informed person who is not close to the field.)

In general there are quite strict deadlines and rules for submission. Any further action after submission, if not explicitly forbidden, is unwise at best. An exception is often made for upgrading information, such as changing "submitted" to "accepted" or "accepted" to "Vol. M, Number N, pp mm-nn." In case of doubt, verify beforehand whether this is permissible and whether such updates must be sent to the committee chairman or secretary (as opposed to being sent to each member).

3.2.2 Closed peer review: Refereeing for journals and granting agencies

The most closed peer review process is the one used for publications in "peer-reviewed" journals and also for some refereed conference proceedings (A and B) and also as an external referee for granting agencies (C).

For all of these (A, B and external referees for C) (i) the referees are chosen from some internal list and perhaps also from a list furnished with the submission, (ii) the identities of the actual referees are always kept from the authors, (iii) referees do not usually know the identity of other referees. (Many journals explicitly ask the referees not to reveal themselves to the authors at any time.) However, as mentioned above, the author can usually request that specific people (such as direct competitors) be excluded from serving as referees, and such requests are almost invariably honored. (The editors/agencies do not welcome possible scandal related to allowing a conflict of interest.) The only real control of the journal and proceedings referee opinions is the editor's (or editorial committee's) judgment. For the granting agencies the evaluation of the external referees' opinions is in the hands of the committee in question. For the journal there may also be a formal appeal process from this first or second round of peer review (involving more referees unknown to you and, perhaps, a known Associate Editor). How can such flawed, complicated and shadowy systems work in the real world? Actually, they work better than you might think.

3.2.3 Closed peer review abuses

Abuses of the peer review system can and do occur, however. (One should expect this, since all human systems are fallible. We do not consider the honest errors where some inferior work slips through and some good work is unfairly rejected. "Abuse" here means that there is some malign or dishonest intent.)

Probably the worst misconduct takes place when a referee abuses the implicit trust (and sometimes the explicit guidelines in a conflict of interest statement) and uses the information received in confidence to gain an unfair advantage. This can include starting or redirecting a competing research program or even holding up acceptance to allow time for the competing program to publish first. All in all, while there will always be more misconduct of this kind than one thinks, this kind of severe damage seems to be sporadic, episodic and fairly rare.

Much more common, but still relatively rare (most referees are relatively honest, and will declare conflicts of interest) is the referee who is familiar with your work or with you, already dislikes the work or you (perhaps because a feeling that you have slighted the referee's work in the past). An ethical but ill-disposed referee should declare this bias (along the lines of, "I am sorry, I really cannot render an impartial opinion here.") and withdraw as a referee. Most referees who are already biased against you will, however, see themselves as noble and unbiased defenders of true science and of innocent journal editors, and thus see no conflict of interest. From this assumed high moral stance they can then proceed to slam you and your work. If you are lucky, this kind of referee might overdo it. This excess may arouse editorial suspicions, whereupon the negative opinion will be devalued, and other opinions will be retained. More subtle and practiced ill-disposed referees will not overstep this line and will thus prove hard to rebut, especially if they avoid too much detail and rely more on adjectival innuendo (e.g., "superficial," "seriously flawed," "slight advance," "there is not enough new science to warrant publication in this journal" and the like).

As mentioned above, if you can become aware of people who are likely to behave in this way, you can ask the editor to exclude some people you name from acting as referees. Do not, however (as we mentioned above), make the list too long, since this will give rise to suspicions of incipient paranoia and perhaps lead to your wishes being ignored.

The opposite case to unfair rejection of papers by journals that should have accepted them is the uncritical acceptance of papers that contain serious flaws, and yet receive the implicit approval of this refereed journal. A common reason for this is that the reviewer(s) are friends of the authors or have such a high respect for their previous work. Hence they read the manuscripts without bothering to provide the constructive criticism that is crucial for the peer review system to work effectively. While this reduces the quality of the publication, it is clearly less unjust than undue suppression of good work. Also, the faults can be addressed by the authors or by others, so generally there is little reaction beyond a shrug of the shoulders. In comparison to the "mortal sin" of being unfairly severe, this is a "venial" (i.e., minor) sin of being too easy.

By the way, by this time you might well be wondering why this potentially dangerous anonymity of journal referees is still accepted. Why not have open refereeing with referees signing their opinions? The reason is pure pragmatism. Without this anonymity the system would grind to a halt. It has been found over the years that if referees (who, one should remember, are unpaid) who give negative verdicts become open to attack by aggrieved authors, they will then refuse to referee in the future. (Some journals ask the referees if they are willing to be thanked by the editor for their assistance in the event of publication. All this does in practice is to allow one to get a partial idea of what set of people are on the editor referee list, but little more. Clearly, for the rejected papers, the referee anonymity still holds, since there is no public place where the referee is thanked for assistance in having a paper rejected.)

The system of anonymous refereeing for peer-review journals is thus unlikely to change in the foreseeable future. (One could call it a Nash equilibrium in the refereeing game — a stable but not totally satisfactory state.) To repeat what was said at the outset, "The peer review system is, like democracy, highly imperfect, but all other systems are disastrously worse."

You may find a lot of this information to be discouraging, and perhaps even disheartening. Most of us go through this disillusion from time to time, tempered mainly by the difficulty of constructing a better system. Science is clearly not well served if you have to struggle to get your work done, because someone else (typically an envious competitor) is trying to trip you up. On the other hand, science is ill-served when poor and even erroneous science is published — but this may well be the same opinion of your detractors concerning your work. The system has to accommodate both points of view to some extent. (It might seem that the best thing to do is to try to become friends with all your competitors, join efforts and do the work together. Unfortunately, this is not always possible. Even with the best will in the world such alliances are inherently unstable.)

In essence, what we have presented here are the rules of the peer review game. If you want to participate, you should at least be aware of them, even if you do not take advantage of them.

In general, if your work is sufficiently important, it will be published even if not in the journal of your first choice or of first rank. If published it will in due time be recognized for what it is, and be copiously cited, even if it was not published in one of the very best, high-profile journals. These "late bloomers" can be identified by citation indices if one is willing to take the trouble. The system does work more or less. "Excellence will out" — eventually.

(An outstanding example of this emergence of an important result from relative obscurity is the work reported by K. Takayanagi and co-workers in *Surface Science* (K. Takayanagi *et al.*, *Surf. Sci.* **164**, 367 (1985)), in which they described for the first time a satisfactory model for the 7×7 reconstruction of the Si(111) surface. This problem had been particularly elusive, and had been at the top of everyone's mind in this field for many years. Although its solution was not published in one of the very top journals like Nature or Science, it has nonetheless been cited more than 600 times by the end of 2004.)

3.3 Ethics in science

3.3.1 Abuses outside peer review

Do not make the mistake of imagining from the foregoing that ethical abuses are confined to peer review. Some of the very worst abuses that occur in the context of publication happen before peer review even begins. While this is a topic that could make a book in itself, the worst abuse is (thankfully) so rare that just a few cases will be mentioned. More common are distortions in the according of credit (usually due to the abuse of power within a scientific team), and these are much more likely dangers to the beginning scientist. (See, e.g., what happened to Federico in Japan, in Chapter 6 "Cautionary Tales.")

3.3.2 Falsifying, "correcting," "discarding anomalous" results or data

This is one of those subjects that could undoubtedly take a whole chapter on its own. The most heinous kind of misconduct as far as publication is concerned is outright faking of experimental data.

If ever you are suspected of or involved in such an affair either directly or as a co-author, *only access to irrefutably primary data can clear you of all doubt.* (This underlines the need to *keep very clear* records and *never to destroy the primary data*.) Sometimes it becomes clear that the falsification must have been carefully planned and that the falsifier cannot have been self-deluded. Sometimes (more tragically) it is a panicky response to time or other pressure to "prove" a theory in which the falsifier genuinely believes, and in the belief that the "right" results will emerge "next time," when the experiments can be "better controlled." In either case, the verdict is the same, in effect excommunication from science for life.

As remarked in C.P. Snow's *"The Search"* (Scribner's, New York, revised edition, 1959), a novel on the effect on a scientist of a doubtful piece of work, "The only ethical principle which has made science possible is that the truth shall be told all the time. If we do not penalize false statements made in error, we open up the way, don't you see, for false statements by intention. And of course a false statement of fact, made deliberately, is the most serious crime a scientist can commit."

By the way, a corollary to all this is that, in a sense, experimentalists are implicitly more scientifically moral than theorists. The argument is simple. In doing theory one is usually working out the implications via mathematics of some assumptions which are explicitly given. There is no room to hide, so to speak. (However when very complicated calculations are done on large computers, because of the complexity of the computer programs and their finite arithmetic, errors can creep in and this kind of work resembles experiment in that aspect. Then this work too becomes open to the abuse and violation of the implicit contract of faith that can occur in reporting "real" experiments.)

In reporting an experiment (or a complicated calculation), while much vital detail is given, much of the standard procedure is taken for granted, including excluding spurious effects, repeating the experiment, not suppressing results which appear "anomalous." This is the implicit bargain in experimental literature. An experimentalist who breaks this bargain, even unknowingly, can have a reputation destroyed. (See C.P. Snow's remarks above.) Even if the error does not become public, the fact that the experiment seems not to be capable of being reproduced will cast a pall over the work. Too many such results and a scientist may find that while the work gets published somewhere, the results will be ignored, because people will distrust them, and will not be interested in taking the trouble to reproduce them. (The subsequent distrusted results will be published, but naturally not in high-impact journals where rejection is easy, but they will be published somewhere because nobody will be prepared to say (even anonomously), "Those results are not real!", implying the author is a knave or a fool.)

3.3.3 Plagiarism

Deliberate plagiarism (among other aspects it may involve misusing privileged access to work as a reviewer or the like) is almost as bad. It is true that the plagiarism can sometimes be unconscious — a half-remembered hint from a forgotten publication of someone else. To avoid the suspicion of plagiarism, the best practice is to do a thorough literature search and then to be very generous in citing the results. It is far better to include very loosely related stuff (with phrases like "somewhat related work has been reported by" etc.). After all, since a deliberate plagiarist would never cite as a source the work actually plagiarized, you are very unlikely to be accused of having plagiarized something from a work you actually cite. Just hope that this work is included in your full round-up of everything vaguely related to your work. When it is a work that was not cited that was apparently plagiarized, claiming that the plagiarism was unconscious is very difficult.

3.3.4 Abuse of power

In the community of scientific peers, some have power over others (often alphas over betas, betas over gammas) and this opens the door to abuse in attributing credit, either giving too much to someone who deserves less or none, or diminishing the credit of those who have contributed significantly. Sometimes it is the contribution of junior members of a collaboration that is being unduly suppressed. Sometimes senior people are added to the author list (perhaps because they help pay for the costs) or because their reputation will help gain acceptance for the work. The abuse of those who are at a disadvantage in power is in effect close to a scientific equivalent of rape or, in less severe cases, harassment. Like cases of harassment, unless the victim is willing to complain publicly, these abuses, like most such abuses of power, are hard to find, hard to nip in the bud and hard to correct. Although some major journals have been trying to obtain statements from collaborations of who did what, the usual result is either silence or an equivalent of "We all contributed something significant, but we choose not to give further details."

If working in industry, there is a likely ethical conflict between "open science results openly arrived at" (to paraphrase Woodrow Wilson) and commercial secrecy "while we'll tell you what our product is (more or less) and what it can do, how we make it remains our property to treat as we like (unless or until we patent it)." In most industrial aspects, as one would expect "money rules" (and does not talk much, at least until the patents are settled). (A particularly sensitive area is drug testing where human lives may be affected, sometimes drastically.)

If working in government, there may be similar problems in publishing results which go against government policy. (Again the most reported cases seem to be those about publication of test results unfavorable to drug companies.)

In both these cases (industry and government) there are usually legal restraints in the form of employment contracts which complicate the ethical situation enormously. Of course with industrial research contracts done in universities, similar restraints apply but the intellectual and legal terrain is treacherous. Intellectual property rights disagreements here can make lawyers very happy.

3.4   When ethics fail

We will only touch briefly on the examples we quote, since they are simply reminders that the topic is important although instances are relatively rare.

3.4.1   A spectacular example of scientific fraud: The Schön affair

In 2002 there were two big scandals related to scientific fraud in physics. In this section, we discuss one of these examples and mention the other.

The summer of 2002 was filled with gossip and controversy as an independent review committee investigated fraud allegations brought by various members of the scientific community against J.H. Schön. At the time, Schön (whose doctorate was in fact then revoked (in 2004) by the University of Constanz in Germany) was a staff scientist at Bell Labs, Lucent Technologies (Murray Hill, NJ). By now, there is in fact an extensive para-scientific literature on this topic, published in various news and commentary features in *Science, Nature, Physics Today, Physics World,* and so on.

Perhaps the most interesting lesson to be learned from sad events such as these is that experimental science is *self-correcting.* If someone were to publish fake data in an obscure journal that nobody ever reads, the practical effect of their misconduct would be close to nil. Since nobody would refer to it or even read it, it would be as if it did not exist. On the other hand, you cannot expect to publish fabricated or falsified data on very hot topics in the best journals, and get away with it for long — it is bound to have such an impact on the community, that people will try to obtain similar results, fail and then "blow the whistle."

Although experimental science is self-correcting however, in this case it has come at a very high price. By the time the fraud was uncovered, many groups worldwide had already invested a significant amount of time and resources trying to reproduce Schön's results. More rigorous procedures and education on ethics of scientific publication may help to avoid this sort of occurrence in the future. In the aftermath of this embarrassing and very damaging episode, Bell Labs introduced a more rigorous internal review procedure, which must be followed before regular submission of a manuscript to a peer-reviewed journal. Internal review procedures, whether formal or informal, actually represent a very good approach which in the long run should improve the quality and rigor of scientific publications.

One of the problems with Schön's story was that he was in the spotlight all the time, because he was working on subjects that are very trendy and considered to be very important for developing new technologies. It is rumored that the management of Bell Labs pushed him forward and continuously encouraged him, regularly issuing press releases on his "fabulous" work. Since he frequently published in high-prestige journals like *Science* and *Nature,* everybody working in the same or related fields was trying to keep up with his work, and many other scientists tried — in vain — to reproduce some of his experiments. This lack of success aroused a lot of suspicion, and ultimately led to the very unpleasant conclusion. In hindsight, what is the most puzzling and unsettling element in this story is the puzzle of how he could think that he could get away with it.

Another issue in the Schön affair which was addressed, but only partially, by the committee is the responsibility of the co-authors. It appears that none of the scientists who collaborated with Schön and co-authored the papers that led to the controversy were even remotely aware of his data fabrication and fraud.

Each of the collaborators brought in a distinctive and complementary expertise, and apparently did not realize what was going on, since they never actually observed him while he was carrying out the experiments (and apparently falsifying or fabricating the data). Obviously this is somewhat related to the current trend of overspecialization in science. However, when collaborating with scientists from other disciplines, it is always advisable to try to learn from each other, possibly doing experiments together or at least witnessing each other's lab work, whenever possible. After all, this is to be considered one of the advantages of working in a multidisciplinary environment — crossing the boundaries of your own discipline to learn something new and therefore (hopefully) exciting.

In an odd coincidence, during the same year another fraud was uncovered, this time in a U.S. government lab where a scientist, one Victor Ninov, a key member of a team which had inferred (through decay product measurements) the discovery of the superheavy element number 118, was shown to have faked the essential data.[b] Again, suspicion arose since no other laboratory could reproduce his results. The results were withdrawn before his data fabrication was exposed after a year's careful investigation and science eventually corrected itself.

---

[b]As reported, for instance, in by Bertram Schwarzchild in the Search and Discovery section of *Physics Today* **55**(9) pp 15-17 September (2002). The fraud extended back to 1999 and investigation started in 2001 and took a year to complete.

Even when isolated quickly, science fraud is a great disgrace for science. There is really no excuse for it. We hope this message is clear enough for aspiring scientists. If you are a good scientist — and there is a lot of evidence to support the notion that Schön and Ninov were technically excellent — you should be patient and wait for good results to make your name known. The easy way will eventually turn against you, and is a very unwise choice. If you are not a good scientist, well, you can certainly find something else to do.

3.4.2 An example of suppression of share of credit

The Appleton-Lassen Equation in ionospheric radiowave physics (with electron temperature and ions neglected) (quoting from http://physics.oulu.fi/fysiikka/oj/761648S/2004-05/Introduction_1.pdf with references omitted here) "was for a long time known as the Appleton-Hartree equation, but more recently the priority has raised some debate. According to a previous common view, the earliest papers containing the dispersion equation were by Appleton (1928) and by Hartree (1929). Appleton's paper is based on a talk given to the International Union of Radio Science in October 1927. It contains only an outline of the theory, whereas the full presentation was given much later (Appleton, 1932). On the other hand, Hartree's formula differs essentially from the Appleton-Lassen result, because Hartree included a Lorentz polarization term in the equation, which is now known to be incorrect. Rawer and Suchy (1976) have pointed out that the correct dispersion equation was really published for the first time in a somewhat different form by Lassen (1927), whose manuscript was received in the journal on 25 July 1927, about three months before Appleton gave his talk. Therefore it seems that the priority actually belongs to Lassen and the equation should be called the Appleton–Lassen formula. This term is used by Budden in his text (Budden, 1985)."

Appleton later received a Nobel Prize in physics largely for his radio wave work, most of it experimental on reflection from the ionosphere. However the basic equation carrying his name (but not that of Lassen) was a foundation stone without which that work could not have been accomplished. It is difficult not to believe that Lassen was ill-used and that lack of generosity was the most charitable interpretation of Appleton's behavior.

3.4.3 A rare example of a deliberate misstatement

In March 1987 a breakthrough article on superconductivity at high temperature appeared in *Physical Review Letters*. Paul Chu, who directed the experiment, purposefully committed an error in his *submitted* manuscript, by exchanging the symbol Yb (Ytterbium) in place of Y (Yttrium) in the chemical composition of his new superconducting material, as a precaution to protect his invention. He then *reintroduced the correct symbol in the proof-correction stage*. What is perhaps most disturbing is that during the peer review process Chu was actually receiving phone calls in which other scientists claimed that the Yb compound did not superconduct, which meant that the content of his manuscript had somehow leaked. [*Histoire et légendes de la supraconduction*, Sven Ortoli, Jean Klein, Calmann-Lévy France 1989].

Chu's suspicion of the refereeing process thus appears to have been justified in this admittedly extremely charged situation.

3.5   Intellectual property rights and patents

If you invent something, you may want to protect your intellectual property. Usually this is done by filing patent applications. If you work for a private company your employer will, typically, be more interested in you either applying for patents (usually owned by the company) or in keeping some things quiet, than in publishing papers in scientific journals. Having students involved can lead to complications, ethical and possibly legal. Which names are on the patent application can give rise to the abuses we discussed before with respect to refereed publications. All this is part of the "intellectual property rights" jungle, feared by scientists, beloved by lawyers.

If you invent a new technique, which may have a commercial value besides its intrinsic scientific or engineering content, you may write a patent, to protect the "intellectual property" of your invention. Some people have indeed become rich from patent royalties, but the hard reality is that most patents do not yield any real money to their holders. This is partly because very often a given market is not ready to launch something so innovative. In fact, it may take more than 20 years for your invention to appear on the market, and by then your patent will have expired. A good example of this is probably the invention of the transistor (Bardeen, Brattain and Shockley, Bell Labs 1948) which appeared on the market a long time after it was first invented (in the early 60's). In fact, from a strictly monetary point of view, you or your institute will have to pay to have the Patent issued and maintained for a certain number of years. It is said that only one patent of every 1000 generates enough revenues to cover its costs, and only one of every 10,000 is actually fruitful and yields real profits to its inventor(s). Realistically, for the individual scientist, it is usually the extra punch that a patent can give to a CV that is important.

In general, you cannot patent a naturally occurring phenomenon (such as the discovery of an element, or a chemical) even if you are the very first person to observe it. Indeed, the prerequisite for protection of your intellectual property rights is that you have *invented* something, and that the invention is not trivial.[c] An invention can be defined in general as "*a product of human ingenuity.*" The concept of patent is there to protect your intellectual property, which implies that you must have *developed* something new, not just discovered something that can be found in Nature.

Patent law may vary greatly from country to country. In North America for example, if you publish a paper that contains the main results of your invention, then you will have one year's time to apply for a patent, following up on that paper. In Europe, by contrast, you have to apply for the patent first, because if you opt to publish first, your patent application will be rejected, on the grounds that your idea will not be considered new any more.

---

[c]In connection with his demonstration of the electromagnetic waves predicted by Maxwell, Heinrich Hertz could not have patented the waves, but he could patent the means to produce them and detect them.

If you work for a university or a Government Laboratory, your best personal option to further your career is to write compelling scientific papers, and, by this disclosure, to let the patent go. Of course in a Government laboratory you still may not have the choice.

DIVERSION In connection with finding a useful application for pure research, we offer the description of another Sidney Harris cartoon, rather easy to imagine, since its caption reads *"Perhaps, Dr. Pavlov, he could be taught to lick envelopes."*

If, on the other hand, you work for the R&D department of a private company, your management (with few notable exceptions) will not encourage you to publish papers or to present your work at conferences, but rather to write patent applications (which of course imply disclosure) or to keep most of your practically important results secret. In fact in industry you go to a conference to learn and not to teach.

Most public scientists think that they would never like to work for a private company. Many are idealists in the sense that while (to be frank) the prime motivation for the work is the fun of doing it, followed by the esteem of one's peers, if benefits are to flow from the work, most would like humanity to benefit, rather than a private company. Those who work for private companies would say that without them those benefits would not flow, because governments are poorly equipped to develop patentable devices to viability in the market, to thus grow the economy and enable (among other things) the funding of more university and government research. Of course, you may well have an opinion which differs from both of these.

Before we leave this subject, it should be noted that the patent system provides interesting niches for those who like science but find themselves in the end not cut out for life on the science frontier.

In the niche we are discussing you can remain in contact with science in your daily practice and put your ability to work in a useful way in understanding science but not innovating in it. This niche is that of a patent attorney or a Patent Officer (one who evaluates Patent applications). Patent Officers are civil servants, with all that that implies in routine and stability. (It might help to remember that Einstein was initially a Patent Examiner and enjoyed the work which left him time and energy to do very successful science. But Einstein's path was not a normal route to scientific fame.)

Successful patent attorneys in private practice make large amounts of money. If that is your goal in life, having a Science degree and competence could thus help you get into that lucrative business. Of course you would have to go to law school (which might open up other career avenues in its own right) and that means a considerable additional investment in money and time, but the results would be worth the investment. In Europe, the European Patent Office (which has three offices, in Munich, The Hague and Berlin) has been hiring steadily for a while now (which means they may saturate their available positions soon).

3.6 Gender-equal opportunity employment

The concept of (Gender-)Equal Opportunity Employer (EOE) goes a long way, and is not of course limited to academic or scientific careers. This is clearly a hot and sensitive topic, particularly in the United States. The topic is one that is difficult to address without stirring up controversy and upsetting people. The intent here is not to address the topic with the weight it deserves, but simply to underline a few points for those who may not have felt the need to face this problem (i.e., many of the men).

Before discussing gender bias in science, it is instructive to consider the fascinating case of the recruitment of musicians for classical orchestras. It had always been easy for the orchestra hierarchy to say that, while they were perfectly prepared to hire women, unfortunately none of them played well enough. It came about that there was a potential problem of nepotism in hiring which was solved by having a double blind audition, where the candidate played behind a curtain without speaking and the identity was not known to the jury. This worked so well that these blind auditions became the norm in North

America. After a while women started emerging as winners once the jury could not see that the player was a woman.[d] Alas, blind auditions do not work for conductors of classical music, who must be seen to be judged, so this would seem to be the explanation for the striking lack of women conductors.

In many science fields, including physics and related specialties, there is ample evidence that the proportion of women representation goes drastically down from female students earning first-level degrees to women who are full professors. This tendency is much less marked in biology, and rather less so in chemistry and astronomy. What accounts for this difference from one field of science to another is still fairly mysterious. Differences between countries are also striking.

There are various plausible candidate reasons for this increasing under-representation. It may be that a significant number of women are simply not interested in pursuing an academic career full time, perhaps because they will give priority to child bearing and rearing. It may be that many women do not want to operate at the extremely high-pressure mode characteristic of the male "high rollers." In medicine, where women have been dominating enrolment for many years, it is clear that even without children a forty- to fifty-hour week is prevailing rather than the sixty-hour weekly schedule which is often the work model for men. In a science department this difference in work effort may make for difficult evaluations. ("Are we judging science output per year or per duty-cycle year?") Given these difficulties in obtaining equitable treatment, some may rather want to work as, say, high school teachers, to have more free time to devote to their hobbies and families. Only the most talented will be able to overcome the barriers in a science world that is largely dominated by men.

Some funding institutions based in several countries have adopted specific schemes to reinstate women into scientific positions, either part time or full time, once they have started a family.

Even were these specific gender biases removed, there is another difficulty which relates having a family pair in which both are

---

[d]Some of this is amusingly recounted by Malcolm Gladwell in *Blink*, Little, Brown (2005).

professionals. As we have mentioned above in considering where to locate, it is clearly far more difficult to contrive to find employment for two people, i.e., the candidate and the candidate's spouse if both are highly specialized. Many universities still have rules against nepotism which make this problem even more difficult.

If you are a young woman and are strongly motivated to pursue a scientific career, you should not let this issue deter you too easily. Rather, you should find a supervisor who is also a mentor figure, and who is willing and capable of supporting you early in your career. It does not matter whether your supervisor is a man or a woman, as long as the person understands the issues at stake, and is willing to help.

In terms of acceptance by the other members of a department in the present climate, however, it may actually prove a good strategy for a woman to find a male supervisor. If so you should probably try to find one who already has demonstrated an appropriate sensitivity to the problem. This might be demonstrated by experience in training several female graduate students, or by his having a spouse, sister or daughter who is pursuing a scientific career.

If your objective is to overcome your own difficulties rather than raise the consciousness of the department for the sisterhood, it would be well to raise the problem early in the process (when it is a future problem) calmly and diplomatically. Excessive complaining and table thumping tends to drive supporters over to the opposition, who already has enough supporters. (As we will see later, in discussing difficult referees, it is a useful tactic to let the opposition seem strident and unreasonable.) The best approach, in our opinion, is to collect sound data on the situation, and to project it to male colleagues as objectively and scientifically as possible, when the appropriate opportunity arises. Do not make it personal or emotional, or you stand to lose. Being males in a department where we have only two female professors, we raised this point already more than once. We did it jokingly rather than emotionally or with a tone of complaint, and we have the feeling that most of our colleagues are beginning to listen.

A few anecdotes (all due to Federico) may not be amiss here.

3.6.1   R., a research associate in biology/pharmacology

Federico: — R. is a research associate in biology/pharmacology in one of the major Canadian universities. We asked her why there are more women in biology departments than in physics departments. She claims with a certain degree of confidence that one of the main reasons why women prefer pursuing a scientific career in the "soft" sciences (such as biology) as opposed to the "harder" sciences (physics, engineering) is that the latter disciplines require a lot more abstract thinking and formal descriptions (in terms of mathematical formulations) than the former. Apparently, a higher percentage of women as opposed to men are turned off by the hard sciences, because of this perception.

3.6.2   J., an engineer

Federico: — Consider, for example, J.'s experience. J. is an undergraduate student at a university in the U.S. and has been working as a contractor for the U.S. military during her time off from school in the summer. I met her at a conference on materials science and engineering. She agreed to be interviewed so that her point of view could be recorded here. She told me upfront that she is somewhat concerned about her career opportunities in science and engineering being hindered by her gender. J. is worried about having to constantly prove her worth in an environment which is largely dominated by men. Since she is also very good-looking, J. is even more concerned about not being taken seriously, or worse, being considered a "blonde airhead."

On the other hand, J. realizes that she may actually have better chances of finding a job than her male counterparts, because of the quotas that are set up in the United States as part of "equal opportunities" and "affirmative action" legislation.

I asked J. if it was important for her to have role models of her same gender around her, and to be mentored by a woman. Surprisingly, her answer was "no." J. explained that she has always been an independent type of person, and therefore does not feel uncomfortable around men. She added, however, that other women who feel more insecure are more likely to feel less comfortable around men, and are therefore more prone to choose female advisors for their graduate or post-doctoral work.

Being a man, I realize that I can hardly relate to what J. told me. I have never been frowned upon (as far as I know) because of my gender. Once however, after being interviewed for a faculty position, I overheard a rumor that the job in question was going to be offered to a woman candidate, because females in that particular department were severely underrepresented. I actually discussed this with a friend, a professor from another university, who cynically commented: "If she is *half* as good as you, she would win hands down."

3.6.3   D. and T. (chemists)

Federico: — At the same conference, I also had the opportunity to interview D. and T., two graduate students in chemistry from a University on the east coast of the U.S. Their advisor is a prominent male professor.

Besides being a young woman, D. is also a foreigner. She moved to North America from Asia several years ago, to pursue her graduate studies. Now that she has almost finished, she hopes to find a job in the U.S., preferably in industry. For her, being a foreigner is an issue she has to contend with, before even bringing in problems related to gender. In looking for a job, D. feels that she is being seen as "different," even though open discrimination is not an issue. In her department she feels all right, but she is not optimistic about finding a job. Generally speaking, D. feels good about equal opportunities and affirmative action programs. She says that because of these programs, the situation in North America is far better than the one in her home country (and in most of Asia).

T. would also like to work in industry after she graduates in a couple of years. Again, because of equal-opportunity legislation, T. feels that in looking for a job she would have a better chance than a male candidate, at least as long as their credentials are equivalent.

T. feels very confident about working in the lab, and yet she does not feel as good about her general knowledge of chemistry and science. Her present advisor is a man, but during her undergraduate studies she had preferred working for a woman. She admits that she looked first at women faculty in her search for a Ph.D. project, but then decided to give priority to the project itself rather than to the gender of her advisor.

T. feels that women tend to be underrepresented in science and engineering for various reasons, including family issues (in our society women are still expected to mostly take charge of child rearing and managing the everyday life of the family) as well as the fact that there are not enough female role models to look up to. As a consequence, women are less motivated to aspire to a higher degree and to a research career in general.

T. did tentatively outline possible solution elements. In particular, she feels it is important to motivate girls and young women towards the study of science early on in their school careers, say in 6th or 7th grade. At the same time, T. feels that the burdens and chores that come with raising a family should be shared more equally between the parents, thus giving women a better chance to pursue their career aspirations, not only in science but in general.

3.6.4 S., Eastern European scientist

Federico: — Culture and politics play a major role especially in relation to issues such as equal opportunities. S. now has a permanent scientist position in a University in her home country in Eastern Europe. She recently returned after working as a Marie Curie Post-doctoral Fellow in Western Europe for about three years. According to S., because of the way of life imposed by the former communist regimes in Eastern Europe, there is virtually no gender issue in science and engineering departments in this part of the world. In fact, S. claims with great confidence that she does not see any advantage or disadvantage in working as a woman in science in her home country.

Chapter 4

Acquiring and Using a Reputation

Sections of this Chapter

4.1 Getting known in your science

Now that we have discussed the game or ecology of science, including peer evaluation and the like, it is appropriate to discuss how to do well in this context. In other words, *"Now that you know the rules, how should you play the game?"* The essential for success is to do the kind of work that can be used to build a winning CV, with appropriate publications, and to become sufficiently well-known to become what the Americans term "a player" (in a positive sense). Then you should capitalize on that investment in terms of improved funding, perhaps of fellowships and of new employment opportunities. This will enable you to produce more publications, and so on upwards and onwards.

In the other analogy we use, assuming that you are a performing artist with enough talent to succeed, what is being discussed here is how, knowing the rules of the game (so to speak) you can be your own agent and tactician advisor.

What will be discussed next, therefore, is the kind of general advice an agent might give on the ways and means of improving your image, on what to do and what not to do and the like. The one topic in that line that will not be discussed in this chapter is the actual writing of publications (as distinct from doing the science), which is of sufficient weight that it will be given a separate chapter to itself (the next, Chapter 5), together with the other less permanent ways of communicating your science, namely, oral presentations and posters.

What we will now take up are, in order, publishing strategy (why and where), conference strategy (why and which), seminars (employment-related and others), employment interviews, and strategies for obtaining funding for yourself (fellowships) and for your research (grant applications).

4.2   Publishing: Where and how

As already discussed, in the medium-to-long term your aim is to build your reputation. The component in this campaign is your CV and its essential core, which is your list of refereed publications.

To make the strongest CV (as discussed in Sec. 3.1) each journal publication in it should be in the highest caliber journal you think best matches the quality and matter of the work to be published. Because others (such as members of evaluation committees of all sorts) may well apply citation indices and impact factors to your list of publications, you should at least be aware of these tools when making your choice of where to publish (see below).

There are other publication opportunities however, and these will be discussed after the refereed publications. Conferences are often a special case, which we discuss at the end of the section (i.e., in Sec. 4.3.8), with a summary here for completeness. If you make a presentation at a conference you will often be pressured into producing something (a sort of mini-paper, usually only 4 pages) for a conference proceedings or the like. To sum up Sec. 4.3.8, in general in our opinion it is usually better to produce a proper refereed publication elsewhere, than to submit the desired mini-paper (unless it is to be published in a high-profile refereed journal).

Yet other publications can be book chapters, or even full monographs. These are sufficiently specialized that there will be no attempt here (or in the rest of this book) to give any counsel on book-writing and the like.

A subject too new to be discussed (in effect the dice are still rolling in this game) is Web publication, apart from the simple placing of copies of your published papers on your Web site as a sort of running CV. (Here you should be aware of copyright issues. While many journals will allow you to post the pdf versions of your papers on the web, many others will not, at least not until you have asked and obtained permission for it.) On the other hand, posting some of your Power Point presentations is perfectly acceptable, since you have not surrendered copyright on these to a publisher.

4.2.1   Refereed journals: A specially important case

We have already discussed peer review and the like in the earlier sections, and the detailed discussion of how to present yourself and your work in a refereed journal will be tackled in the next Chapter. Hence the only topics discussed here are measurement instruments used by "bean-counters," deans, managers and the like. There is a regrettable tendency nowadays to quantify scientists' and institutions' worth by calculating their impact as derived from citation indices, and considering also the impact factors of the journals in which they publish. Although the aim is to provide evaluation tools to goad scientists towards a higher quality, this is a dangerous game, since the indicators being used are not always objective and reliable. Nonetheless, on the principle of knowing the tactics used by others, we turn next to discuss impact factors, citation indices and so on, so you can see how your work appears to people using these tools.

4.2.2   Citation indexes and impact factors

The impact factor is generally defined as the total number of citations of papers published in a given journal over the last two years, divided by the total number of papers published in that journal in those same two years. In essence, it represents the average number of citations expected for an article appearing in that journal, at least over a given two-year period. Although it may be thought of as a relatively useful indicator of the quality of a journal, this concept is unfortunately widely abused. In essence its statistical significance is very limited, because the spread around this average number of citations can be huge. (A rarely-cited paper in a high-impact journal may actually be better than a similar paper in a journal of lower impact, simply because it was likely refereed with much more care because of the fierce competition to be published there.)

Sadly, many funding agencies and even universities across the world are now turning to what they call "quantitative, objective indicators" of a scientist's performance, and the impact factor is often taken as one of such indicators.[a]

Clearly, scientists whose work is never cited should seriously consider doing something different with their lives. However, publishing in the very best journals is basically a lottery, even for the best scientists, and therefore a lot of people decide that it is not worth their time and effort to seek glorious publications, for example in *Science* and in *Nature*.[b] They simply submit to good journals in their field of research. This, however, tends to restrict their audience, and consequently their work is less cited and less known.

These scientists do good, honest work, and in our opinion they should not be penalized for their choice of not competing aggressively to publish in flashy journals. In earlier times when the concept of impact factor had not been invented, scientists did not have to worry about all this. They would publish their work in the best journals of their field, and that would usually be enough to achieve the recognition they deserved and needed.

---

[a]A new scientist index, the *h-index* is reported in Nature v.436, 900 August 8, (2005), an invention of Jorge Hirsch (see the original article at arXiv:physics/0508025 v4 23 Aug 2005). It reports on obtaining a Hirsch index h or h-index h (an integer) where you have published h papers referred to more than h times and the other papers are referred to less than h times.

[b]The impact factor of *Nature* has recently hit 30.979 (just beating out *Science* at 29.162), an unprecedented value. The journal reported this success on its cover in 2004, adding pompously that "No *Nature*, no impact."

Citation indexes are undoubtedly a useful working tool, however their present use to evaluate scientific performance on the part of managers and administrators has become, in our opinion a dangerous and counterproductive way of "doing business." However, faced with this tendency, you have two choices; either play the game or suffer the consequences.

4.2.3 Citations: Strategies and consequences

At the end of a scientific contribution there are — almost always — references to previous work. It is extremely rare to read papers that do not refer to other papers, because nowadays it is not easy to say something that is 100% new. (Even if you were to say something totally new, there is always a context for the work, so it is always desirable to refer to other work. You should define the context of your own contribution — what was the state of the art before you published, and how your work advances the understanding of a certain problem. In addition, as remarked above, a partial but effective defense against accusations of plagiarism is to cite pretty well all the relevant literature.)

To put it somewhat brutally, one can say that citations are to be thought of as an exchange currency. If you cite other people's work (whether they *really deserve* to be cited or not is, cynically speaking, a different issue altogether) they will be happy.[c] This is a good way to make friends in your community. It should not have escaped your notice that if you cite someone in your work they will be more likely to read it more carefully and more likely to cite you in their own work. Paying attention to how your work may be cited and by whom is not just a matter of ego. Nowadays having many citations is — perhaps sadly — a measure of scientific value (read the section about impact factors) and may therefore be decisive in getting your research funded, or in securing a bonus or even a promotion. Thus, having many citations to your work

---

[c]Another convenient Shakespeare (Hamlet) extract on how to treat the traveling players puts it neatly. Polonius: "My lord, I will use them according to their desert." Hamlet: "God's bodykins, man, much better: use every man after his desert, and who should 'scape whipping? Use them after your own honour and dignity: the less they deserve, the more merit is in your bounty."

to show (for example when applying for a promotion) has become a matter of survival in today's scientific world. Of course, this does not mean you should cite someone's work even if it is of poor quality, just to make new friends. But it does no real harm to cite good work which is a bit peripheral.

On the other hand, if you forget to acknowledge your predecessors and their achievements, either intentionally or accidentally, this is one of the best and fastest ways to make enemies in your field of research. (In an elaborate spoof on a typical string theory paper, a pseudo-reference was produced along the lines of "6. At this point we should refer to the work of an eminent physicist. Since he won't refer to our work, however, we won't refer to his.") Also, as mentioned above, the best way to stop a charge of plagiarism is to cite anything which might be considered a source.

Federico: — As an example, take R.'s experience with citations. She was once browsing an issue of a medium-level journal in the fields of surfaces and thin films. (By "medium-level" is meant that it is a journal well-used by the specialists, which contains the occasional excellent and important paper.) R. stumbled onto an article from an Asian group, whose members had been studying a topic (the hetero-epitaxial growth of semiconductors) which was precisely the subject of her Ph.D. thesis. The authors were reporting a slightly different approach, and obtained results similar to hers. R. read the paper very thoroughly and critically, because it described a system that she knew very well. She found several inconsistencies in the paper, and, what was worse, the authors had not cited her work. R. wrote a polite e-mail to the corresponding author of this paper, pointing out that she had been working on the same topic, and attaching her own work as pdf files. The senior author of the paper replied cordially, thanking R. for her criticism and for sharing her work.

R. then discussed the issue with a colleague/friend, who knew this group because they had carried out part of their experiments in his Laboratory at an International Facility. Together, R. and her friend decided to write a Comment on this article.

R. wrote to the Editor of the journal just after she had started drafting the Comment, and asked him if the journal accepted Comments on work previously published. After consulting the other Editors, he replied that they had decided to accept Comments.[d]

When the senior author got the Comment from the Editor, who asked if there was to be a Reply, there was considerable discontent. Nobody likes to have their work publicly criticized. (By the way, this is another reason to write good papers. It is not wise to write something defective, because someone may well get upset about it and write a nasty comment on that defect, which, if justified, can be a real slap in your face. Besides hurting your ego, it will also (negatively) affect your reputation.) The result (as often happens) was an aggressive Reply, in which the original work was strongly defended together with an attempt to discredit R.'s criticism.

This is not an uncommon result, but it can sometimes be avoided by a conciliatory approach seeking a common publication for a correction or amendment to the previous work. (Clearly science is best served by this approach, since there is a consensus. Unfortunately many letter journals will only allow such joint publications as Errata (which are usually misprints and the like), while Comments invariably indicate an adversarial approach with significant content. In a perfect world Amendments or Clarifications with significant content (and perhaps co-authored with the original authors) other than Errata would be allowed.)

There was only one issue without controversy in the Comment/Reply exchange. A few months later, R. decided to try a series of experiments to address this particular issue, so she took the unusual step of writing back to her competitors and asking if they wanted to participate. R. wrote a polite letter to the senior author, explaining that she had no hard feelings about their previous exchange and proposing that they should get together and collaborate on this experiment. The competitors gladly accepted, and now they are very good friends and have even started to exchange visits between their respective laboratories.

---

[d]He also said that it would have been the first time that a Comment was published, and that they had decided it would be good to accept Comments because it would imply that at least someone had read a paper in the journal and had taken it seriously.

The moral of all this is that if you cite generously (and avoid cutting remarks in so doing), you can expect your colleagues and even your competitors (within reason) to do the same with your work, and to be your "friends." On the other hand, if you are stingy in your reference list you are definitely heading for trouble and incurring unnecessary resentment.

4.3   Conferences: What they can do for you

In general, the best way to present your results is to publish them in written form, since *"Verba volant, scripta manent."* ("Speech is fleeting, the written word endures.") As we keep reminding you, writing compelling papers for peer-reviewed journals is undoubtedly one of the very best ways (perhaps the very best) to advance your career and make a mark in your field of research. Given this, why should you go to a conference and present your results more directly, it is true, but in a way that is much more ephemeral?

(While most conferences are published in some form, those results are often extremely difficult to find in a few years, unless the results appear in a peer-refereed journal. The modern tendency to distribute CD's is worse because they are likely to be much harder to find than a book (which might be in some library) for someone who didn't actually attend the conference and receive the CD as a result.)

The advantage is that you gain a brief opportunity to make a strong impression on those in the audience (or those who go to your poster). The analogy here might be to an author on a book promotion tour. (The author is promoting the book; you are promoting your work.)

There are four main reasons to attend a conference beside the simple act of presentation itself. These reasons can be caricatured as (1) Advertisement and Feedback, (2) Gathering Intelligence, (3) General Networking and (4) Exploring Opportunities Elsewhere. Let us take them in turn.

4.3.1 Conference reason #1

Advertisement and Feedback. You want to present your results to an audience that you hope may be interested in what you have to say (and perhaps in following up the details in print), and you hope to get some feedback from them.

As well as these laudable aims, you also want to remind your fellow scientists of the written work you have published, draw their attention to your latest results (as yet unpublished) and to the future perspectives of what you are doing. Basically, the underlying aim of your presentation is to convince your colleagues that what you are doing is interesting and promising, and that they should definitely find the time to read your recent papers in the literature, and look out for your future ones (or ask for preprints). If you are convincing enough, some of them may even propose to collaborate with you (however no more than half of them will prove to be serious about it). Do not forget to bring along several reprints of your recent papers as well as a stack of business cards, to keep in touch with your new acquaintances. (Bring business cards even if you are still a student — they will also make you look a lot more professional to your senior colleagues.) It is true that with e-mail available, many now prefer to be sent the material electronically, but others may want to read such material on the way home (or even at the conference, if particularly interested). This, then, is the usual primary and explicit reason for attending a conference.

4.3.2 Conference reason #2

Gathering intelligence. At a conference you will usually hear the latest and most exciting news in your field of research. (It actually depends considerably on the conference; unfortunately some meetings afford quite poor pickings.)

In terms of intelligence gathering, remember that the papers published in journals online or in print *this* week contain information that was reported at conferences *between three and six months ago.* Conversely, the results presented at a conference *this week* contain results that will be published within the *next three to six months.* If you

want to keep abreast of the events in your field, you should plan to attend
at least one conference per year, possibly two (also depending on how
fast your field is moving, and how fast you move relative to it).

4.3.3   Conference reason #3

General Networking. Conferences are great venues for *networking,*
the discussion and negotiation activity necessary to put in place the
informal arrangements that enable much research. Meeting other
scientists in a relaxed atmosphere and discussing many topics informally
is a great way to make friends and to sound people out. Since the
research core in most fields of research tends to be relatively small,
making friends is vital. Ultimately your success will depend greatly on
how many friends you have, and who they are.

By the way, for the young scientist, here is a necessary word of
caution. Many scientists (and especially famous ones, or the others who
think they are or *should be* famous) tend to have a *BIG* ego. This often
means that they are easy to offend, even if that is furthest from your
mind. (An interesting definition of good manners: *"A person who has
good manners never gives offense unintentionally."*) Since it is risky
to offend such people (for example if you are invading their territory),
if you know that Dr. Famous may be in the audience, listening to
your talk,[e] you should certainly consider citing Dr. Famous during your
presentation, perhaps referring to Dr. Famous' seminal work in this field,
which has "… inspired you and many other colleagues to do *etc. etc.*"
(Of course, you should really do this only if the work is actually relevant
in the context of what you are presenting, otherwise you will sound a bit
ridiculous if not downright silly.) It is better if you actually have that
written down somewhere conveniently on one of your images, so that
you do not forget to say it, and so that it is visible besides being audible.
Even if Dr. Famous has already left (perhaps his talk was scheduled

---

[e]This is somewhat unlikely, since most of these big shots tend to show up 10 minutes
before their talk, give their performance — lousy performances are not as uncommon as
you would hope — and then leave about 10 minutes after their talk is over, hinting that
they are going to another meeting or anyway that they have something better to do after
having listened to themselves.

before yours, or perhaps he did not show up at all, which is another prerogative of famous, arrogant scientists), his students or post-docs or friends may be in the audience. So you should still watch what you say, and also what you don't say! You may find that this type of over-diplomatic conduct is distasteful. In some ways, we agree with you. However you have to live in the real world of scientific research, and this simple advice may turn out to be useful. (Perhaps, if *YOU* happen to have a big ego, or develop one, there will come a time when *YOU* will get upset if others do not appreciate and appraise your work!)

There is one more aspect that we have hinted at but which we bring to your attention in connection with highly developed egos, and that aspect is *territoriality* which is usually expressed in *turf wars*. Because of his paper(s) on a given topic in the past, Dr. Famous may come to be convinced that this is his "turf," so that you should not blunder into it without his blessing. If you find this out early enough, Dr. Famous can usually be disarmed by asking about this work, whether it was followed up or not. If you do not cite him at all you may expect trouble. (A mentor may be able to provide valuable guidance here.) This applies even more if Dr. Famous is in your own department, since there is often an unwritten law (to avoid turf wars on one's doorstep) that researchers do not trespass on the research areas of other members of the same department except when involved in a specific collaboration project between them. Think carefully before you enter the forbidden garden, and estimate the cost beforehand!

4.3.4 *Conference reason #4*

Exploring Opportunities Elsewhere. Another excellent reason to attend a conference is for checking out job opportunities, or even posting your own job advertisement (although arguably this is again part of the networking motivation) or looking for an extended visit or even a sabbatical somewhere else. In general, the "big" conferences (in the materials sciences they would be the APS, MRS, AVS, ACS meetings in the U.S., EPS, ECOSS etc. in Europe, and so on) will offer an employment center, with companies, universities and government labs posting their ads and sometimes even carrying out on site interviews. In smaller conferences you must shift for yourself.

4.3.5   Go if several reasons apply

Generally speaking, the best conference situation is when several of these reasons apply. For example, if you cannot present anything "hot" (e.g. you have just started a new project and do not have enough new results yet), or if the conference is in a field which overlaps only partially with yours, or if the people you want to network with are unlikely to be present at the meeting, perhaps you are better off saving your time and money for the next available conference that meets these requirements. There are, after all, a great number of scientific meetings every year, so it will be relatively easy to find a new one which is a better match to your requirements.

4.3.6   Invited talks at conferences

An important exception to this rule of thumb (i.e., of going only when more than one condition is satisfied), is when you have an *invited talk*. In that case you should most probably go, *even if the other reasons are not met*. It is relatively rare to receive an invitation to give an invited lecture at a conference, and if you get it, you should take advantage of it. Especially at an early stage of your career, an invited talk at an international conference really stands out on your CV. It shows that your peers think particularly highly of you, and that your work is having an impact in your field. It is the sort of recognition that boosts your confidence and visibility, and is therefore very important. Perhaps one day when you are famous you will turn down some invitations to talk at conferences, or generously pass them to your post-docs and younger collaborators. Until then, you should accept them all, even if the organizers do not offer to contribute to your travel expenses.[f] You should

---

[f]Until about two decades ago, in most cases an invitation to give a lecture at a conference would be accompanied by an offer to reimburse all, or at least most, of the invitee's expenses. Sadly, this is rarely the case now. Funds to organize conferences are becoming more and more scarce, partly because too many competing conferences are being organized. Also, meetings organized by large organizations tend to be more expensive and even more stingy, because a significant part of the revenues from the conference will be used to pay the salary and benefits of the organization's employees.

consider such expenses as a wise investment on your CV, and therefore on your future.

4.3.7   Conference trip funding

Going to a conference has to be feasible within some travel budget and must be a justifiable expense in terms of that source of funding. If your travel is controlled by your supervisor then the rationale for going to the conference must be thus justified and negotiated between the two of you.

Before asking your supervisor to send you to a given meeting, it would be wise to come up with a realistic budget. (Your supervisor will probably ask you to do that anyway, so it will look better if you do it without being asked.) For example, you can check out the best airfares on the internet; however your most important task is to check whether the conference offers a discounted registration rate for students (sometimes even student grants), and other similar discounts, like cheap student accommodation. (You may benefit from knowing that many conferences actually hold prize competitions for the best student oral and/or poster presentations.) Once you have come up with a total budget, it will be much easier to determine whether there is enough money for you to attend the conference. If other group members are looking into attending the same conference, you may look into a collective budget.

Traveling to conferences and meetings (which are occasionally held in beautiful locations) is perhaps one of the most interesting and appealing fringe benefits of our job. The temptation to do it often is therefore quite strong. Unless you are famous and are invited everywhere (and have *unlimited* funds to spend on travel), however, you should not exaggerate. In fact, after a few meetings, you are likely to start hearing the same talks and results over and over again, in which case you are better off staying in your lab or at your computer and actually trying to do something new.

4.3.8   Conference papers (proceedings)

Presentations at conferences often have a hard-copy after-effect. If the upshot is a refereed paper in a respected journal (typically this is only for Invited Presentations) then you may treat this as a reasonable publication like any other. (As an example, Invited Papers at the annual meeting of the Division of Plasma Physics are usually published (after refereeing) in a special number of Physics of Plasmas. In terms of other papers in Physics of Plasmas these are more highly rated because the selection process is fairly rigorous — albeit somewhat political.)

Some conferences (often in Europe, or involving a considerable third-world representation) attempt to pressure authors of submitted contributions into submitting a (four-page) paper based on their refereed abstract, which then comes out (nowadays) on a Compact Disc, which only the conference attendees can obtain. In days gone by, a conference volume or two was produced instead of the CD. Years after it may be quite difficult to find a copy of the volume in question. (One reason for this state of quasi-publication is that conferences attended by scientists from the developing world often find that these people can only attend if the conference is "refereed" and a polite fiction is maintained that a paper might be rejected, even if they never are in practice.) Clearly this limited distribution is hardly a publication in the strictest sense of unlimited life in accessible form. For this reason many people tend to ignore the strictures on redundant publication and publish essentially the same material (often mixed with other stuff) in a proper refereed journal where future access is more or less guaranteed. This of course is in conflict with most journals strictures on prior publication of the work being submitted, and further devalues the conference quasi-publication. However this adulteration is still a fact of life.

Unless you are at a very early stage in your career and desperately *need* to publish extra papers, it is best not to provide contributions for conference proceedings. These contributions are rarely seriously peer-reviewed, and this is reflected in the quality of the published papers,[g]

---

[g]One notable exception that comes to mind are the proceedings from the IEEE series. Those tend to be peer-reviewed quite rigorously, thus keeping a high standard.

which are often not nearly as good as the peer-reviewed papers that appear in regular scientific journals. In any case, few scientists today have the time to read conference proceedings. Papers that appear in proceedings volumes are rarely cited, since they are not read in the first place. Your peers' opinion of you will depend on how much and what you publish, and how often your work is cited. In the short run, when you are a student and your publication list is short, your CV may benefit from that extra paper, even in a proceedings publication. If this is the situation, you should discuss this option with your advisor, on a case-by-case basis. In the long run however, publishing an extra paper, that is probably a re-cooked version of something you already published, and that will not be read and cited, is not a good strategy, since it diminishes your overall impact. To make your mark as a scientist, your best strategy is to do good work and publish in terms of quality, not of quantity. As an indirect benefit to the environment, avoiding the publication of inferior papers will actually save some trees.

Indeed if the results you present at the conference are good and original, you should write them up and submit them as a regular article in a peer-reviewed journal. If they are not that good, perhaps because they are preliminary and/or incomplete, you should wait until you have a coherent story to tell. While preliminary results are generally accepted at meetings in oral or poster form, they are not well received in print (what you publish in written form is meant to last, to "stand the test of time"). Also, double or triple publications with very similar titles and contents are not well viewed by other scientists. The scientific literature is already clogged with papers, many of them being fairly useless. Thus contributing "cloned" papers will not win you any credit or "brownie points" with your peers or your administration. Remember, in this job your peers' opinion of you is critically important for your well being. Perhaps the best compromise is to cite the work presented at a conference as an un-refereed talk/poster. Not furnishing the four-page paper for the CD makes this a reality.

There is one last aspect about conference proceedings. They usually come with a deadline to submit your manuscript, more often than not during the conference itself. This can be good or bad, depending on your personality. Some people work effectively only if they are under pressure, for example to meet a certain deadline, and probably for them it is good. (If you want to know more about the advice on this specific topic, read the section on *Meeting deadlines*.) Other people do not like to be pressured. Rather, they like to take their time (sometimes in a glacial scale) and think their material through thoroughly and carefully. One might say that in the long run this latter category tends to produce the best science (with some exceptions, of course). On the down side, these practitioners of the "few but ripe" school of publication also run the risk of being scooped, which is not a pleasant experience. You should find out early on which category you belong to, and govern yourself accordingly.

4.3.9   Posters or oral presentations?

When discussing posters (which are also discussed in Sec. 5.6), we were not able to come to a "we" consensus between us, so we decided to present each point of view separately. The difference is probably due to the difference in the way the important meetings in our respective fields are organized. We each naturally tend to support what works for us in our discipline.

Federico: — With very few exceptions, I personally think that posters should only be used for a small meeting. When going to a major conference, one of your main goals in presenting your results is to increase your *signal-to-noise ratio*. (Your signal is represented by your own work, and the noise level is what everybody else is talking about.) This means essentially that you want your work to *stand out*.

At a *large* meeting with a great number of attendees, the only way to be visible is to have at least an Oral Presentation (This can be called *parallel* visibility, since you reach out to everybody at once.) If you have only a poster (this is *serial* visibility), there is the chance that nobody will show up to look at it, or perhaps only your friends will come over. This is frustrating on one hand, and on the other hand it means that you are wasting your time and money.

At a *small* meeting, conversely, it may be good to present a poster. Small meetings tend to be more relaxed and informal, and you are not competing with hundreds of other people to present your work. It is not uncommon to have lively scientific discussions in front of a poster, and this is particularly true at small workshops and conferences.

Tudor: — If your presentation is an *Invited Presentation* it will always be an oral presentation, so the question of a choice does not then arise. (However sometimes the speaker will be requested to post the images used in the talk (without staying by the poster) for subsequent study by those who could not be at the oral presentation. This is also handy for people to leave requests for copies on the poster sign-up request sheet.)

For *contributed* papers, on the other hand, the choice between these two ephemeral modes of presentation should be determined by the nature and practice of the presentations at the conference, and in particular by whether the conference is dominated by many sessions in parallel (a likely event if the conference is very large) and whether the time limit for contributed talks is short (15 minutes is common).

If the meeting is *small*, say less than 200 attendees, with no parallel sessions, the choice can be a matter of simple preference. Most would opt for an oral presentation which will be heard by most attendees. In such meetings, however, there is usually a limit on the number of oral presentations (including "one to a customer") so you may be forced to do one or more posters in any case. (A small meeting with an evening poster session with beer and wine can be most enjoyable!)

If the conference is *large* it is likely to be dominated by many sessions in parallel, with a severe time limit for contributed talks (15 minutes is common). The decisive point is then whether your type of subject will be accorded a session of its own.

If your topic has a session of its own, it is in effect a small mini-conference, the people who are interested in subjects related to yours will probably be in attendance at that session, and so an oral presentation is a natural and excellent choice. (You are only tied to your talk for a limited time, and your responses to any questions are available to all in the room.)

If, however, the sessions are fairly heterogeneous, or if your subject does not (yet) fall into a major theme, and you present in a very mixed session, your best friends and important colleagues may well miss your isolated talk, especially if a speaker does not turn up and the chairman just keeps going, so that your talk is ahead of time and likely to be missed by your friends and allies.

In the case that you are likely to be in a heterogeneous oral session, it is best to opt for a poster in a poster session with like-minded neighbors. It may also be true that people in your topic have become poster people at that meeting, in which case you do likewise and join the crowd. When you give a poster it is true that you are tied to the poster for a time much longer than the 15 minutes or so for a talk (it helps of course if you can time-share with co-authors), but detailed exploration is easy for the visiting experts in a way that is impossible for an oral presentation , and of course you will have a sign-up sheet for requests. (Responding to these is now very easy by e-mail.)

4.4   Seminars: What they can do for you

Arising out of conferences, or in parallel with travel to a conference, or through personal contacts, you may arrange or be asked to give a *seminar* on your work at an institute or university department or the like. By *seminar* is meant simply a talk before a group of modest size (at a university department or a research institute) at a reasonable length (somewhat under an hour is usual). Thus you can explain your work at ease with more time than allowed for all but the most exalted of invited talks at a conference.

Apart from the pleasure and luxury of explaining your work to people who want to hear about it, not to mention the fact that it looks good on the CV, seminars can do a lot for you, if handled well. The most important seminars are those given in connection with an application for employment, and are included therefore as part of the job interview. The more informal seminars (usually via an invitation from a colleague) are an excellent way to deepen useful contacts, perhaps a first step to arranging for a collaboration, or possibly for a sabbatical visit (later on). Another important aspect of a seminar is to help in finding and evaluating students or post-doctoral fellows.

Although a seminar is a form of oral presentation, just like the ones you give at conferences, it should be a quite different *performance*. Intrinsically different from conference presentations; seminars often tend to be a lot more informal, and the audience is frequently much smaller. (But again, that may depend on the size of the conference, and on the size of the department where you are invited to give a seminar.) At the same time, the duration of a seminar is typically 40-45 minutes, and during this time you are expected to give a more general introduction and more comprehensive description of your field of research. This is your chance to tell a *complete* story, and you should take full advantage of it.

A more sophisticated form of seminar is called "*colloquium.*" Typically a "normal" seminar is given to a *restricted* audience within a given department (e.g. the astrophysics community in a physics department, or the organic chemistry community in a chemistry department). A colloquium on the other hand is meant to be more general, with a very broad scope, and is thus intended for the whole department and sometimes even for people in other departments. (Nobel Laureates typically tend to attract a crowd.)

4.5 Employment interviews

To respond to a job announcement, you must usually tailor your CV with respect to the specific advertisement to which you are responding. Typical ads specify the set of skills and experience they are looking for, and you should make sure that you are emphasizing these aspects appropriately in your CV and possibly also in a cover letter, in which you describe concisely who you are and why you are the best candidate for the job.

If you are applying for a post-doctoral position, this material (plus reference letters which are usually sent separately) is normally enough. If, on the other hand, you are applying for a faculty position, you are expected to submit also a statement of teaching philosophy and a description of research interests (often also with a fairly detailed request of start-up funds). A job ad for a position in industry or in a government lab is likely to demand other sets of skills.

The burden is always on you to demonstrate, first on paper and then in person at the interview, that you are the best person for that job. In this sense, the importance of writing a good cover letter to outline your overall skills, competence and fit with the job profile cannot be overemphasized. This document is the "set-up" for the interview, and you will have succeeded if much of the interview time is spent asking you to expand on what you raised in your application.

If your CV elicits enough interest, you will typically be invited to an interview "in house", usually with a seminar to deliver (preferably before the interview). This formal interview is sometimes preceded by a telephone interview, which may or may not represent a screening process in itself.

To help to understand the university hierarchy when navigating your approach, the next rather long *DIVERSION* may be of some help.

DIVERSION on the University Hierarchy. (Apologies to the creator of Superman, as heard on the radio "Leaps tall buildings in a single bound! More powerful than a locomotive! Faster than a speeding bullet..."). The source is ANONYMOUS: **Dean:** "Leaps tall buildings in a single bound! More powerful than a locomotive! Faster than a speeding bullet! Walks on water! Gives policy to God! **The Department Head:** Leaps short buildings in a single bound. More powerful than a switch engine. As fast as a speeding bullet. Walks on water if sea is calm. Talks with God. **Full Professor:** Leaps short buildings with a running start and favorable winds. Almost as powerful as a switch engine. Faster than most BB pellets. Walks on water in an indoor swimming pool. Talks with God if special request is approved.

Associate Professor: Barely clears a garden shed. Loses tug-of-war with a locomotive. Can fire a speeding bullet. Swims well. Has been talked to by God. **Assistant Professor:** Makes high marks on walls trying to leap tall buildings. Is run over by locomotives. Can sometimes handle a gun without inflicting self-injury. Talks to animals. **Graduate student:** Runs into buildings. Recognizes locomotives two times out of three. Is not issued ammunition. Can stay afloat with a life jacket. **Undergraduate:** Falls over doorstep when trying to enter building. Says, "Look at the choo-choo." Wets himself with water pistol. Plays in mud puddles. Mumbles to himself.

Department Secretary: Lifts buildings and walks UNDER them! Kicks locomotives off the tracks! Catches speeding bullets in teeth and eats them! FREEZES water with a SINGLE GLANCE! IS GOD!!!

In preparing for your interview trip, you should focus on the seminar you will have to give, and on any other type of presentation that may be required of you. (Some liberal arts colleges in North America for example will ask you to prepare a full lecture (at the undergraduate level) on a given topic. These presentations are extremely important, since they are seen as a sample of your lecturing and teaching skills, and you should take them very seriously.) Remember, this is a competition for *one position*, so it is not enough to be good: you have to show that you are the *best*, or you will simply not get the job offer. Here you really have to optimize and maximize your signal-to-noise ratio with respect to the other candidates (whom you will rarely know).

Good advice on interviews can be found in other publications e.g. Feibelman[2], and the article by Matt Anderson "So You Want to be a Professor?" pp. 50-54 in the April 2001 issue of *Physics Today*.

ANOTHER DIVERSION Collective nouns on campus:
An Arrogance of Deans, A Complaisance of Professors, An Ambition of Associate Professors,
A Jitter of Assistant Professors,
A Bewilderment of Instructors,
A Hunger of Part-Timers,
A Starvation of Teaching Assistants,
ALTOGETHER A Paranoia of Faculty.
From the Chronicle of Higher Education v. 19, n. 2 (14 January 1980) from *More Random Walks in Science*[1]

4.5.1 Preparing for the job interview

After the seminar that allows you to show your talents to the institution in general, there is usually an interview with the selection committee associated with the position for which you have applied. It is your CV (and perhaps some people that you have met) that have resulted in the job interview. This job interview is particularly critical, since it will address other aspects besides your science accomplishments and ability. It is *essential* to prepare very seriously for this interview.

Think of yourself as a chess player in a match game. You prepare for the general lines of attack to be expected and also for the variations in play. As in chess, this is "over the board" and is played over a limited time.

The difficulties and challenges you can encounter during this interview may vary enormously from place to place. You are likely to be severely questioned on almost everything you say if you are invited for an interview from one of the top schools in the country. (For example, if you are seeking employment in the United States, the top schools would be Harvard, MIT, Caltech, Princeton and so on.) You may get some tricky questions even over a meal, during which you are more likely to lower your guard and be less critical and judicious in your replies. At smaller schools the committee may well be friendlier, and will probably have fewer expectations. Such schools do not typically get line-ups of star candidates, and their selection criteria are less stringent.

In all cases, however, during this interview you should expect to be asked (among many other things) what your immediate research plans are, how much they will cost, and how you plan to secure the necessary funding once your start-up funding runs out.

You should therefore do the following things to prepare for all this in the actual interview.

(i) *Prepare in advance what you have to say about your research plans and the like*, almost as if it were a second seminar (also because some departments will ask you to describe your plans during your seminar to the faculty); if you prepare for it, your discourse will sound a lot more fluent and professional, and this will typically make a good impression on the committee.

(ii) *Write out in detail what your start-up funds requirements are.* Bring a copy for each member of the committee (and some spares), and be prepared to justify each line of your budget, should they question you about it. It may be wise — depending on where you are being interviewed — to ask for a little more than what you really need, so that you have some margin of negotiation. You certainly do not want to take a position where they will not be giving you enough funds to start a successful research program;

(iii) Try to find out in advance who are the members of the selection committee, and study in detail some of their more important publications before you go there. It is also wise to prepare some questions on their recent work, since it will make them feel good about it in front of their colleagues. (Note how this discussion parallels that for Dr. Famous at conferences. In effect there may well be several Dr. Famous figures on the selection committee.) If possible, draw comparisons between your work and theirs, and indicate how your expertise may nicely complement theirs so that they may be acquiring a bright collaborator and team player, not simply a colleague. Whenever possible, use psychology to your advantage, and try to avoid surprises. Any surprise/unexpected question may be fatal to you, despite your ability to improvise.

The best approach to prepare for an interview itself is to try to put yourself in the interviewer's position. In many ways, an interview is a psychological game, even like a game of chess. You want to make your interviewer feel at ease, and that hiring you will be the right choice. What skills and personality traits will be of interest to this particular employer? That is why it is important to do your homework about the prospective employer *before* you show up for the interview.

Show your interest and target it to each particular employer at the start, when you prepare your application material. (Note that, to be able to build this rapport, there must be enough overlap of interest between what you do and the work done at a given department. If not, then perhaps you should not even apply there. In the remote case in which they were to offer you a job, if you were to accept you would likely end up as an isolated misfit.)

Think things out in advance, and try to be a good chess player. If you are well prepared, it will be much easier to get an offer when you go to an interview.

4.5.2 During the job interview

Although this will not be easy, during an interview you should try to stay inwardly relaxed. You should make an outward show of serene confidence (without appearing arrogant). However you actually might feel during your interview day(s), it is important that you project an image of success, of a constructive attitude (suppressing any hint of irritation or frustration) and of strong determination. You want your interviewers to have positive thoughts of you when they write their reviews about your visit after you have left. The impression you want to leave them with is that it would be great to have you as a colleague, and that your presence would make their workplace more interesting and, if at all possible, bring direct benefits to them and to their own work. If they turn you down, you want them to at least feel regret that there is a candidate who, although less appealing than you, fits the job description much better.

It is very important that you are interactive and constructive during an interview. You must be at the same time a good listener, but you should also ask the committee members pointed, intelligent questions. The best way to do this, as suggested previously, is to do your homework (thoroughly!) about the department you are visiting, and its inhabitants.

If you look nervous, although it is a perfectly normal and human reaction, it will be noted by your interviewers, and it will probably not score points in your favor. You should answer all the questions you are asked, trying to be as brief and concise as possible — get to the point, don't beat around the bush. Interviewers do not like candidates who talk too little, or too much. At the same time, make sure *you* ask a fair amount of questions. The questions you ask (in principle) serve a double purpose: getting information that is relevant to you, and possibly making you look/sound smart.

In fact, an interview is a situation in which double feedback should be exchanged. As much as the prospective employer needs to find out enough to decide whether to hire or not, you will want to learn enough to figure out if you really want to work there or not. (After all, up to this point you are going on external information and what you have been able to dig up for yourself.) In principle the first issue is the more important one. (Clearly, if you do not convince the selection committee, it does not then matter whether you would have taken the job or not.) However, a good interviewer will try to please you as much as you try to please the interviewer. Remember this in the future when you are on the other side of the fence, acting as interviewer!

4.5.3 *Travel costs for the job interview*

First, if you do not have the means to finance the trip (because it is not a research expense) you should make sure that your travel expenses will be properly reimbursed. This is often the case, but regrettably, not always.

Federico: — While I was about to finish my Ph.D., I was once offered 50% reimbursement to go to the Netherlands for an interview for a post-doctoral position. The person I was talking to argued that I could easily find some other nearby Institute where I could give a talk and thus get the remaining 50% reimbursed. Needless to say, I did not even remotely consider doing all this.

In my opinion, if a prospective employer does not offer full reimbursement of your travel costs, there are very good chances that you are not being taken seriously. Unless you are desperately seeking employment under a severe time limit, I strongly advise against this type of compromise. You should have more respect for yourself. In any case, if your prospective employer is not taking you seriously, you will probably NOT get a job offer in the aftermath of your trip. (This may be because more competitive candidates have also applied, so you are being kept as a convenient backup.) Even if you do get the job, this cut-price attitude may well indicate that you may not have your boss' full respect. There is also the possibility that your interviewer is simply stingy, which is another good reason to avoid the job, since this attitude on spending money would later affect your daily life in the laboratory, most likely in a negative way.

At the same time, if you are desperately in need of a job but cannot find one in a reasonable time frame, perhaps you should seriously consider the possibility of actually changing careers altogether. Although it may be a disappointing realization, it is possible that you are not suited for a scientific career (and *vice versa*), and in this case you should change course immediately.

4.5.4   Mock job interviews: Two examples of failed interviews

In this section, we describe two examples of *failed* interviews, i.e. situations in which the candidates *did not* get a job offer. Although these examples are obviously not encouraging, you should learn from them and not let them discourage you. After all, you should always remember that the majority of job interviews *do not* lead to a job offer, but not necessarily because the candidate "blew" the job interview.

4.5.4.1   C.'s experience in North America

C. was invited to an interview in a very prestigious University in North America. He had joined the faculty of a smaller University just the previous year, but was facing some problems in adjusting to his department's philosophy, and decided to go for another round of applications, to see if he could do better.

During the interview, he was eager to show some recent results he had acquired during an experimental shift at a synchrotron radiation facility. He had used a new technique, and this had permitted him to shed new light on a problem he had previously worked on.

Unfortunately, one of the members of the selection committee was an expert in this particular technique, and started grilling C. about the technique itself. Of course C. knew the basics about the technique, and had rehearsed what he knew for the interview.

However he had never intended to become an expert in that particular technique, but rather had just used it once, by way of a collaboration with local experts at the synchrotron, to study the problem he was interested in. C. is a *problem-oriented* scientist: in developing his research program, he identifies a topic, then uses all the techniques that can help him understand the problem at hand. The interviewer, on the other hand, is a *technique-oriented* person. He tends to develop his research program around a specific technique, asking himself "What can I study with this technique?" He was keen to show off his expertise to the other members of the committee, perhaps even saw this as an opportunity, and did not mind at all that it was entirely at C.'s expense.

C. tried to defend himself, describing what he knew about the technique and pointing out that for him it did not make sense to learn so much about it, since he did not plan to make it his main experimental tool. However, his approach was not well received by the committee. C. did not get a job offer in the aftermath of his interview trip, and later found out that this occurred precisely because of this unfortunate episode with the interviewer who grilled him.

To the selection committee, C. simply appeared to be too eager to show his latest results, without having understood in sufficient detail how he had acquired them. Had he been aware of the interviewer's obsession he probably could have easily avoided confrontation, being suitably deferential to the interviewer's expertise and emerged a winner. In effect, he trod on a buried land-mine (but perhaps one that might have been detected by studying the publications of that member of the Search Committee).

4.5.4.2   F.'s experience in Europe

F. is a native European and had been through a series of job interviews for faculty positions in North America. His next one was in Germany, and this was particularly important to him because if he had succeeded, he would have been able to stay in Europe. Unfortunately, when preparing for this particular interview he forgot to take into account the cultural differences between the two continents, and basically pitched the same story that he had been selling in North America. In retrospect, that was clearly a huge mistake.

In North America, a university that is hiring wants to hear a proposition for an ambitious research program, in which the new faculty member will be the lead investigator. In Europe on the other hand, departments want to hire someone who will help build on current expertise, in a very collegial and collaborative way. (We are not saying that North American departments frown upon collegiality and cooperation; but at first, particularly during the tenure-track period, they want you to develop something completely on your own, and prove that you are a truly independent scientist.)

F. greatly overdid this aggressively independent aspect during his interview in Germany. His flamboyant, ambitious "American-style" presentation was not well received, and F. did not get a job offer as a result of his trip.

4.6 Getting your science funded

When you begin your evolution towards becoming a more or less autonomous scientist, your awareness of funding aspects is pretty well at the level of "Will my boss pay for this?" It is only natural that in the early years you worry about the work and leave someone else to worry about the money.

Since "He who pays the piper calls the tune," the person who has to be convinced so that you can pursue a given line of research, buy a piece of equipment, pay for a big repair or a trip is the one who is paying the piper. When it is your boss this is often done quite informally, without even the ritual of a handshake. The skill that you need to develop then was how to deal with this person — "how to play the payer," so to speak.

A useful image is that of a bird in a nest, where the food for the nestlings is brought to the nest by the parents, and the essential skill for this nestling is how to convince the parent bird to feed this nestling rather than the competing siblings. When the nest is left, the young bird now has to become an independent forager, and the new foraging skills must be quickly learned.

If you are working in an *industrial laboratory*, in effect you remain in a (larger) nest, and management will generally provide the money you need to do your research. Of course, as the "payer," it will call the tune and tell you as the "piper" which projects to tackle. (There are a few exceptions to this. For example, IBM, HP and Bell Labs (Lucent

Technologies) still offer the possibility to carry out basic research at competitive levels.)

DIVERSION *Judging Research.* "View from the Bottom" G.E.K. Mees (Eastman Kodak research head for many years.) "The best man to decide what research work shall be done is the man who is doing the research, and the next best person is the head of the department who knows all about the subject and the work; after that you leave the field of the best people and start on increasingly worse groups, the first of these being the research director, who is probably wrong more than half the time; and then a committee, which is wrong most of the time; and finally a committee of vice presidents of the company, which is wrong all the time." *More Random Walks in Science*[1]

Of course one can be sardonic on the micromanaging of research in industry.

As one should expect, Sidney Harris has another classic cartoon on this aspect. The caption is what the manager is telling his research team, *"Due to a tightening of the budget, we are forced to curtail our overtime and weekend schedule, and request that all major breakthroughs be achieved as early in the week as possible."*

It is up to you to decide whether you like these conditions or not. In *A Ph.D. Is Not Enough*, Peter Feibelman[2] describes very clearly the advantages and disadvantages of working in a *managed* environment (i.e. *industrial or government laboratory*), so this topic is not repeated here. It is best to decide early on whether this life suits you, because making the transition to the life of an independent university researcher is much harder unless your industrial work has made you a potential "star" in academia with a reputation strong enough to afford substantial funding. It is true that in industry there may be work which is supported by outside grants and in that case you may be called on to play a role in the preparation of these grant proposals. The procedure varies too much from one research environment to another to discuss the details, but some profit can be obtained by reading the material below on applying for research grants. We suspect that most of the people who are reading this will probably have opted for the freedom of working in an academic setting, so we will concentrate on that aspect.

While you are developing your career skills, you must develop the techniques to get your research funded from various sources by applying for money in writing.

However, until you have some faculty standing or the equivalent, you will not be eligible to apply for research grants yourself, although you may be asked to write some paragraphs for a grant proposal. If such an opportunity comes your way, seize it enthusiastically. It is an excellent way of seeing how the process works and the experience should serve you when you will be applying for your own grants.

What you can often do officially is to apply, not for project funding but for some sort of *fellowship* to *support yourself*. Since some aspects of writing a fellowship application are very similar to the same aspects in applying for a grant, it is convenient to discuss them together. After all, in applying for either a research grant or a scholarly fellowship, you are trying to convince a funding agency that you and your ideas are worth investing on. The sooner you learn how to do this effectively, the better.

4.6.1   Fellowships and grants: Common elements

The usual way for a student or post-doctoral associate to begin is to apply for a fellowship from some funding agency. Most fellowships will only cover your salary[h]; however even if only the salary is covered they offer at least three intrinsic advantages:

---

[h] There are a few exceptions, including the following: (i) Marie Curie Individual Fellowships offered by the European Union, which also offer a small travel budget; (ii) NATO fellowships offer relocation benefits; (iii) Humboldt fellowships offer various benefits.

(i) Since your supervisor now does not have to cover *your salary*, more money will be presumably available for *your project*, and to cover other expenses. It is also possible that a generous advisor may actually increase or supplement your salary (in those institutions/countries where this is allowed), and use the rest to send you to more conferences and do more things. Thus both you and your advisor will be happier and overall more productive.

(ii) If you have a fellowship before arrival, you can pretty much choose where to go to work. Clearly, if you are depending on someone else to cover all your expenses including your salary, your choices will be severely limited. However your freedom of choice may be much more limited than you might expect, since many funding agencies ask you to choose where to go at the time of application, and may not offer the possibility of changing destinations later. You should verify just how portable your fellowship will be if you succeed. You should also verify whether your future employer would take you if your fellowship application did not succeed. In any case, even if your employer chooses to accept you without knowing if you are bringing a fellowship with you, if the fellowship is accorded you will have your employer's respect and gratitude, and are likely to be tangibly rewarded by the employer.

(iii) There is a large common element between the application for a fellowship and a grant proposal. In both you are trying to convince a funding agency to give you money to do something that you believe is scientifically important and which you will be able to do. If you get a fellowship or two early in your career, when you apply to become a junior professor later on, you will have already established a track record in bringing in money, and applying for grants will be somewhat easier. It will certainly give you an edge on competitors who either did not take the trouble to apply for a fellowship, or were not good enough to receive fellowships.

Sometimes, however, the timing will just not work out. Most funding agencies accept applications for fellowships only once or twice per year, and this timing may not coincide with your personal schedule in looking for a job. There is no simple way around this, except to emphasize that, for any application, whether for fellowships or grants, you should always be well aware of the deadlines, and should do as much planning as possible so as to avoid unpleasant situations later on. While on timing, it is essential to start in time to do an excellent job on any application. Ideally, you should write the whole thing out, beginning, say, *at least two months* before the submission deadline, and then put the whole thing in a drawer to "cool off" for a week or so. Then pull it out, read it with a cool and skeptical eye to give it that final polish, and send it off in plenty of time. If an application is worth making at all, it is worth making in the best way possible.

As pointed out in preparing for an interview, accurate planning, access to information, avoiding surprises, and being a good chess player will give you an edge over your competitors and a better chance to succeed. This applies here too.

One of the tricky aspects of getting a fellowship is gaining the necessary knowledge about the scholarship programs that are offered, their deadlines, and finding out if you are eligible.

Federico: — As a graduate student, I spent a significant amount of time looking into such programs. Without any specific guidance, it was already clear that fellowships = opportunities in a scientist's early career.

In today's modern, global world, it is actually not very difficult to find this type of information, especially with a few hints. A lot of the information for physicists is contained in ads that are listed in monthly publications, like *Nature, Science, Physics World, Physics Today, Materials Research Bulletin, Chemical and Engineering News* and so on. (Again, scientists from other disciplines will hopefully be kind enough to excuse us for not knowing the equivalent publications in their fields; suggestions are in fact welcome for future editions.) These journals are discussed somewhat in the Sec. 2.16 on *"Keep yourself up to date."* Information on other programs can be found for example by browsing the internet, or from hearsay. Even after the student/postdoctoral phase it is worth spending a significant fraction of your time performing these searches, not because they will be of immediate use, but because this will be useful to your students or to someone else you might know.

Applying for a fellowship and for a general-purpose grant have sections that resemble each other in that you will have to describe the salient features of your research either since you began or over some period specified in the grant application rules.

You may be asked for a summary of your best work and of its impact. You will also be required to indicate a research plan of some sort. It helps a great deal for all this if you have a well-developed CV that has been up-dated regularly. It is a matter for reflective judgment just how much "hype" you should put into the description of your work and plans. Too much and you appear as a callow, shrill and insecure salesperson for your science, too little and you may appear not to have a high opinion of your own work. ("He that bloweth not his own trumpet, his trumpet shall not be blown." From *Littlewood's Miscellany*, Cambridge University Press.) Here is where consultation with a mentor or a trusted colleague can be extremely valuable.

4.6.2 Fellowships

As we just discussed in the previous section, a small fraction of top students and post-doctoral fellows may receive direct personal funding in the form of a fellowship, or scholarship. This funding usually covers most or all of the person's salary and sometimes also some extras for travel and perhaps supplies for experimental laboratory projects.

Such fellowships tend to be extremely competitive, and usually only the top students are able to win them. Salaries awarded through fellowships are often more generous than the ones offered by supervisors from their research grants, partly because they are meant to reward the very best. This is another reason why they are sought after so much. Also, long after the fellowship money is gone, the presence of the fellowship in your CV is a permanent benefit.

When you apply for your first fellowship as a student, for example for an M.Sc. or Ph.D. scholarship, since your experience in research is usually limited, great importance and weight are given to your academic performance, i.e. your grades. At later stages in your career other criteria (e.g. your publication record) will become more important. Typically you are expected to supply a (fairly detailed) project in your application. In most cases you will have to "negotiate" this project with a prospective supervisor, in accordance with his overall research program and in relation to how your interests fit with his.

Applying for a fellowship typically represents your very first contact with a funding agency. As such, it will help you build your track record with respect to funding, and is therefore to be considered a very useful exercise. As you will find later when you come to apply for research support, you must apply in time, apply often, keep on applying and do not become despondent when you are refused. Your lack of success is not a matter of public record, so it is really only your bruised ego that requires toughening. As in any sport, the ball will not go in the goal if you do not shoot, and in most sports (not basketball), it takes many shots to get one goal. You must learn that this heartache of a rejection is part of the "cost of doing business" (at least until you get your Nobel Prize).

4.6.3   *Grant applications/proposals*

As remarked above (but is being repeated here to emphasize the importance of this point), give yourself adequate time to prepare your grant application. Sometimes you will need collaborators (likely from other universities, research laboratories and industry). This may require approvals from the hierarchy in those other institutions, and may take precious time. All this is much easier to accomplish if there is not undue time pressure.

Besides the science part of the grant application or grant proposals, which is the subject of this subsection, there are many important proposal-specific aspects with which we do not deal. While much can be learned by getting people to show you their successful grant applications to use as models, there are useful texts which can help and which are worth consulting.[4]

Contrary to most young scientists' expectations (a lot of people just consider it a boring exercise), writing the science part of a grant application (the part that is most fun to do) is not enormously different from writing a scientific article. The main difference, in general, is that when you write the article you already know the results, while in the grant you are indicating what you hope they will be. One of the things that you must remember for a grant application is that it will be looked at by experts (essentially referees) who will either write reports or be members of the evaluation committee. Obviously it is vital that these experts be convinced.

However, as remarked above, there will be on the committee also some *quasi-experts*, people who have some knowledge of the field, but not an expert's knowledge. (They are, however, the experts in other fields.) It is very important to convince these quasi-experts by explaining the important stuff to them often in a carefully crafted summary sentence or two, of exemplary clarity, usually at the end of the very technical sections. You want to leave the impression that the quasi-expert figured it out without the subtle help you are providing. If well enough done these readers will almost feel that they invented (or "would have invented") the basic concept themselves, and will become partisans as a result. Often you will be asked for a popular summary, perhaps for a possible press release. Your chances of ever seeing this as a press release in actual print are very slight, but you should nonetheless seize the opportunity to explain the overall thrust of what you are doing with a minimum of jargon. If you can be clear there, the reader will be more inclined to give you the benefit of the doubt on the more complex stuff. The point is that here you will not be viewed as talking down to the committee (whom you do not want to offend) but around them to the public. This effort will also stand you in good stead when you have to explain what you are doing in a very limited time (for example to visitors).

DIVERSION Proposals by mathematicians are reputed to be the sketchiest of all. (Mathematicians are often considered to be very arrogant and above such things; they of course simply say that they have a just opinion of their own worth.) One outstanding mathematician's statement of work was simply "I will continue to study the work of Ahlfors." A project officer claimed that many proposals from Harvard or Princeton or MIT were pretty well, "Here's my address. Where's my check?" (From A *Mathematical Apocrypha*[1])

Although the funding agency's aim is to invest the taxpayer's money on the best projects, it is not very reasonable for them to expect (as they often seem to do) that scientists will actually discover *exactly* what they propose to do in their grant proposals. After all, scientific *research* (as opposed to *development*) is about doing something really *new*. When you set out on a new project, to some extent you actually expect and hope to be surprised, and in many ways you welcome the unexpected. If you already knew what you are going to discover, it would hardly be research at all! As Neils Bohr is said to have remarked, "Prediction is difficult, especially about the future."

In this connection, if you have been considering the future, it is not however uncommon to know at least some of the results already, even when you are writing a grant proposal on a topic that is supposedly new.

There is a deceptive way to reduce the risk somewhat. If you can do it, it is an excellent strategy (for a sufficiently productive scientist) not to publish all the results immediately, but to keep some (good) results in the drawer. If you are in this happy position, when you come to write your grant proposal, you can then describe with reasonable certainty what is going to happen, and you can even produce some preliminary results (taking them out of your drawer), to give the impression that what you are proposing is promising and feasible. That way it will be much easier to obtain the grant, since your proposal will look more convincing and realistic. (Of course it will also be much easier to report on it once you are finished, since part of the work is already done.)

You must, however, be careful and emphasize that what you are proposing is "new", by which you mean (which is true) that it has never been reported before. Of course, at the same time, you must make sure that nobody is competing on the same topic, because you certainly do not want to be scooped on results that you are holding in reserve. For this reason most scientists early in their career tend to feel that they must put it all "in the shop window." This strategy of holding some results in reserve for the next proposal is thus likely to be useful only somewhat later in one's research career.

Many funding agencies also ask you to describe what are the "benefits" or "added value" to your country, should the project be funded. This is a very tricky section, because the risk is that funding agencies in time will come to fund only applied research (on the grounds that basic research does not carry enough "added value"). This section may thus prove to be the most difficult one in the grant, especially if you are setting out to do something fundamental rather than applied.

There is no easy way out of it, and what can be said is that it should be taken very seriously and will require your best thought. You must find the right balance between "promising the moon" and being demolished by experts who know better. Discuss your ideas with some colleagues, and try to be creative. If there are only long-term benefits, be honest about it. In academic research, the most tangible benefit to the country doing the funding of a given project is usually its contribution to the training of "highly qualified personnel," who are then expected to join the workforce and make a difference with the specialized training they have received. Emphasizing this point is always a good idea.

Other funding agencies with a specific agenda (e.g. the military), while perhaps inclined to fund basic research, will want to know what are the benefits with respect to their specific goals.

The funding agency to whom you are applying will usually ask you to report on your results periodically and often (for a project grant that is not continued for many years) a final report on how you spent the taxpayer's money at the end of your grant. A successful grant application has usually offered a road map for the research and what will be required in a report is how the journey went, how far you progressed, and if there were any extra unlooked-for windfalls. Many funding agencies are flexible, but even they are usually happier if you tell them that you managed to do what you had envisaged or promised in your original proposal.[i] (However some funding agencies are not flexible at all and may be ill at ease with windfalls. You should check before you send in the report.)

---

[i] It remains a mystery how a funding agency can ask you to submit an original proposal, based on new ideas, and then expect you to deliver exactly that once your grant has expired and you are asked to report. They seem not to understand that research is about the unknown, and that surprises are to be expected.

4.7 Information is essential!

As we hinted in the previous section, while we were discussing fellowships and how to get the relevant information on scholarship programs, information is essential in our modern, global society. In fact, a recent Nobel Prize in Economics was awarded for studies on markets governed by asymmetric information.

There are many ways to obtain information, including browsing the internet (a favorite modality of today's students), reading, and talking to your peers and exchanging information. There is, however, such a significant amount of very useful information out there, that you may want to exploit but which may not be efficiently advertised. This naturally brings us to the next aspect of information, which is *filtering* it. Every day we are bombarded by information of various types, starting from news broadcasts for example. It is essential therefore to *filter* this information efficiently, and determine what you are interested in, and how you could exploit it. This is by no means an easy task, and yet it is essential, because amassing huge amounts of information is basically useless if you cannot use it properly.

In a world where intelligence may cost human life, having access to the correct information (possibly with time to spare) is absolutely vital. Information in a scientist's career means opportunities of various types, and therefore gathering information and filtering it appropriately is a very important task that you will have to manage if you want to successfully pursue a scientific career.

Nearly all of the material on sources was obtained by simply surfing the web and looking into a few key journals. (See the lists after the end of Chapter 7.) Finding the material is not difficult if one uses the power of the Web intelligently. The filtering and selection of the really good stuff will take considerably more time than you would expect, but it will be worth it.

Chapter 5

Communicating Your Science

Sections of this Chapter

As we have said earlier, while the research itself is up to you, in the world of peer-reviewed research, research hardly exists until it appears in the "literature," by which is here always taken to mean the *peer-reviewed* literature.

To get to this desired state, the report of the work must first pass through the "gatekeepers," these being the editor(s), and then the referees. In effect the report of the work must sell the gatekeepers on its quality and originality and on the clarity of the report itself. As we have said before in connection with grant applications and the like, your chance of success will be better (not to mention the quality of the communication) if you take care to appeal to two classes of people. One class is made up of people who have a superficial knowledge of the field (the editor and the browsing reader) who must, so to speak be wooed. The other people are the experts (the referees, the authorities in the field and your critics), who know the field but must be convinced that your work is worthy by the standards of the field. Similar concepts apply to the doctoral thesis, except that in that case the public also consists of the examiners who must be convinced and of the people who need the detailed information in the thesis but which may not be in the publications which should flow from it.

Before and after this archival peer-reviewed output there lie the more ephemeral (but nonetheless extremely important) communications delivered directly to the public in the form of oral presentations, invited talks, seminars and conference posters.

In a much more restricted format, but no less important are the *curriculum vitae*, the traditional way to communicate your worth and provide the links to your work for such vital aspects as employment, fellowships, scholarships, prizes and the like.

These are all ways which you should master to communicate your science to the various publics, and these are the topics of this chapter.

(Although we had planned to discuss here the details of how to write applications for Scholarships, Fellowships and Funding in general, on looking at what was already written for these cases, we found that what we wanted to say was previously handled in the Chapter on getting your research funded. Instead of paraphrasing that information here, we ask you to refer to that material in Sec. 4.6.)

5.1 Scientific writing: Generalities

Being a good writer is important. As a scientist, you want to be a good communicator, and to divulge your ideas widely. A good scientist is expected to communicate results and conclusions effectively, in writing and by direct presentations, both to an audience of scientist specialists from different fields and to the general public.[a] This ability distinguishes (at least partly) very good and good scientists from the average or below.

Beyond this somewhat platitudinous view of the public communicator, there is the fact that, as indicated in previous chapters, the scientist who wishes to succeed in science must be able to communicate

---

[a]There is no need, however to go to the lengths depicted by the Sidney Harris cartoon, in which the text begins, 'CHAPTER 7. THE STRUCTURE OF THE NUCLEUS. "What?" exclaimed Roger, as Karen rolled over on the bed and rested her warm body against his. "I know that some nuclei are spherical and some are ellipsoidal, but where did you find out that some fluctuate in between?"...'

at many levels in science, to peers (and through anonymous peer reviewers) to peer-reviewed publications, to funding agencies and to various committees (again through peers) for funding and academic recognition. Most of this communication is written, at arms' length, so to speak, when you are not present, and effective written communication becomes essential to success.

This is not the place to learn the basics of prose writing in science.[5] (There are many books for that[5]!) The only thing that must be kept in mind is the central goal is clarity; there should be no doubt as to the meaning of any sentence. Rather, this is the place to discuss how to package and color the messages you want to send, to realize that you will always be sending more than one message at a time, and to understand and control all the messages that you are sending.

Your most important underlying message, one which you cannot avoid sending, is the one of who you are, or at least how you appear. Since you cannot avoid broadcasting some message of who you are, you must learn to broadcast the message that you choose and not a worse one by default. In the musical My Fair Lady, the linguist Henry Higgins proclaims that "The moment one Englishman opens his mouth, he makes another Englishman despise him."[b]

The Canadian media guru Marshal McLuhan has also proclaimed that "the medium is the message." It is equally true that often "the medium is the messenger too" or, perhaps, "the message is the messenger." Anything of any length that you write shows something of who you are. However, like an actor in a play, if you pay attention to how you write, you can learn to appear to be something better, and even become so by practicing hard at the appearance. Be aware that your voice will be in your prose and try to step back from the work and see what kind of a person you would seem to be.[c]

---

[b]The origin of this is actually Shaw himself in the preface to his play "Pygmalion" (the source for "My Fair Lady"), "It is impossible for an Englishman to open his mouth without making some other Englishman hate or despise him."
[c]Richard Rhodes (author of, among other things the Pulitzer Prize-winning "The Making of the Atomic Bomb") in his fascinating little book "How to Write" (William Morrow, New York (1995)) has a very perceptive chapter on "Voices." Among other gems he quotes Ralph Waldo Emerson, "A man cannot utter two or three sentences without disclosing to intelligent ears precisely where he stands in life and thought ..."

Another aspect that you can learn to keep in mind is that usually you are engaged in advocacy — you are putting forth a point of view and trying to get the reader to agree. The more the readers value the person you seem to be, the more likely you are to convince them.

Structuring a text can be done much more effectively if you imagine a rather skeptical reader and answer questions which such a reader might well come up with.[d] It is even better if, in the text, these questions can be answered before the reader thinks of them. If you are successful the reader will begin to think that "You know, this author is really quite intelligent and someone to get to know." This is natural, because this feeling implies that "This author thinks as I do and is thus worth listening to."

As we repeat through this book, you should try to impress two levels of readers.

One is the eagle-eyed professional, perfectly at home in the discipline, an expert Doubting Thomas. It is invaluable if you can persuade a colleague to perform this function — that of the Devil's advocate — by an almost hostile reading before documents are sent out.

You should also, however, try to communicate through the text with someone like an informed layman, perhaps another scientist not at all in your specialty, perhaps even further away. Here again, for really important documents it is worthwhile testing the text on a colleague who is not too close to your work. (Some of the top-ranked journals include this sort of intelligibility for the non-specialist in their criteria for acceptance. They know well that good scientists like to graze a bit outside their specialty and this wider circle of readers will increase the journal's impact.)

As we have stated before, in connection with grant applications and the like, in any committee of your peers this targeting of two levels of reader is often vital. There should of course be an expert or two in your domain, but there will usually be many more who could easily understand the work if it (or at least the principal points) is simply and

---

[d]The classic example of this is Steven Weinberg's 1977 popularization of cosmology *The First Three Minutes* (Basic Books), where he says that the book is aimed at a skeptical and shrewd lawyer, one who knows no mathematics but is able to follow an argument closely.

clearly explained. Most members of such committees like to believe that they are not narrow specialists and can get the gist of most things that come before the committee. If you can clearly explain the essentials such people will be much more inclined to accept that you know what you are talking about in the difficult and abstruse sections that they do not really follow. They will feel that much better because they are finding themselves able to follow something noticeably outside their area of expertise, and their opinion of your work will likely be improved considerably. These people also vote on decisions and can sometimes counteract the excessively hostile expert. You may even find that the expert will approve of the way that you can summarize the core of your work and infer that you are thinking clearly and are thus less likely to go astray.

It is true that, in aiming at two publics in the same document, the result may be a bit uneven, stylistically speaking, with dense and complicated paragraphs made up of long and complicated sentences (often so because of length limitations) and much technical verbiage being interspersed with shorter paragraphs, with short, clear sentences with little technical jargon. If this is the price of clarity and of being able to address a wider public, then so be it. Clarity and breadth of impact are worth the price.

(If you are sufficiently successful in science you may be called upon to produce a popularization for the general public. At this point the only respect to be paid to the expert is to avoid saying anything actually technically incorrect, to which one can point and say, "That is clearly wrong." What you strive for in the popular presentation is (as always) clarity. Decide exactly what and how much to say. Better less and clear than more and overdense. If a technical word must be used, define it. This is all that we will say on popularization.)

The order with which writing topics will be treated in the rest of this section is the order in which the young scientist might be expected to have to come to grips with them. This order is (as given above) *Peer Reviewed Publication* (5.2), *Theses* (5.3), and *Curriculum Vitae* (5.4) (Recall that the other important components for which writing skills are required, namely, scholarship and fellowship applications and research proposals, have been dealt with in Chapter 4.)

Books (apart from chapters contributed to edited compilations) are such a large topic and one which does not usually come up early in a scientist's career that we have decided not to discuss it here. (Perhaps we might do it in a second edition, if there is one.) Again we cite without graphics the Sidney Harris[1] cartoon showing Professor Hamlin on the telephone exclaiming incredulously, "You mean *Casey's* book on Hamlin's Syndrome will be out before *my* book on Hamlin's Syndrome?"

The communication topics which are not just a written text are *Oral presentation and organization* (5.5) and *Poster organization and presentation* (5.6), which are treated for convenience in the last two sections. (Of course we are aware that placing these last two sections (on communication by something other than text alone) in a logically distinct place, is not the order in which they are required, since the beginning researcher may well have to use these oral skills long before being faced with writing something significant for publication.)

5.2 Peer-reviewed publication

As indicated earlier, next to obtaining the results which are the object of your work, publishing good papers in scientific journals is probably the most important single task you should be performing to advance in your career (this of course presumes that your results are well worth publishing). Trying to build a career without these fundamental building blocks is next to impossible.

We will now turn to discussing some particular aspects of peer-reviewed publication. As we do so, it is worth emphasizing the fact that, the more senior you become, the longer the time you will spend writing. As your career progresses, you will spend less and less time in the laboratory, with more and more time directing those who do and advocating for the work thus done (not only in peer-reviewed papers but also on many other levels). The effort in improving your writing skills for peer publication will be invaluable in these other areas as well, and we will be turning to these areas after we have dealt with peer-reviewed publication.

By what is now a pretty well unshakeable tradition and by the relentless pressure to save space in scientific journals, the scientific publication is about the scientific results and not at all about the details as to how they were obtained. Early scientists like Johannes Kepler or William Harvey would often describe in detail their voyage of discovery (which indeed could help in convincing the reader). By the time of the mathematician Carl Gauss (who wrote originally in Latin), the tendency to conceal (to an almost perverse degree) how the ideas arose became dominant.[e] Like Lieutenant Joe Friday on the old show "Dragnet" on US television, all that is to be communicated are "the facts, ma'am, just the facts." As Richard Feynman observed on his Nobel Physics Prize address (1966) "We have a habit in writing articles published in scientific journals to make the work as finished as possible, to cover up all the tracks, to not worry about the blind alleys or describe how you had the wrong idea first, and so on. So there isn't any place to publish, in a dignified manner, what you actually did in order to get to do the work."

Science is to be communicated in a fashion which resembles the way that mathematicians communicate their mathematics in print, and not at all how they communicate with each other in the conversation in the corridor. You only get to tell the full story if you get a major prize and thus obtain a license to expand and put flesh on the bare bones of the refereed publication. Perhaps in the future with the huge resources of the Web we may be allowed the space to fill in these details in a non-refereed appendix to which the reader could gain access which would not be part of the refereed "just the facts, ma'am" literature, but that time is not yet.

As we have said before, we do not wish to overlap with the usual texts[6, 7] on how to write science papers and the like, but we cannot resist the temptation to point the reader to some satirical "guides" as well, which go under various names: *"Do-it-yourself CERN Courier writing kit"* (CERN Courier July p. 211 (1969), see also *More Random Walks in Science*,[1] p. 140, *A glossary for research reports* in *Metal Progress* v.71

---

[e]See P.J. Davis and R. Hersch's (1981) classic *The Mathematical Experience*, now in paperback (Mariner, Houghton-Mifflin).

p. 75 (1957) *A conference glossary* on p. 173 of *Proceedings of the Chemical Society* (1960) see also *More Random Walks in Science,* p. 167-168. In the kit there are four tables of phrases which can be combined on the principle of (in order) any one from table A through D in succession to give such gems as "Presuming the validity of the present approximation ... pursuit of a Nobel prize ... will sadly mean the end of ... the future of physics in Europe." *The conference glossary* is a translation guide: e.g., in a paper "Preliminary experiments have shown that ..." really means "We did it once and couldn't repeat it ..." in an oral presentation "Why do you believe ...?" really means "You're out of your mind!" *A glossary for research reports* is in a similar vein: "... of great theoretical and practical importance" really means "... interesting to me ...," "Presumably at longer times ..." means "I didn't take the trouble to find out," "While it has not been possible to provide definite answers to these questions ..." really means "The experiment didn't work out, but I figured I could at least get a publication out of it." The lesson here is to look and see to what extent your prose is subject to this kind of cynical misinterpretation.

5.2.1 Letters vs. regular papers

In general, the normal means of publication is the peer-reviewed scientific paper. Shorter publications (Research Notes, Brief Communications and the like) are *either* for more limited topics not up the weight of a regular paper — snippets, if you will — *or* for brief letter-length reports on very important topics for which rapid publication before a wide audience is deemed essential — like *STOP PRESS* bulletins. It is the usual assumption that this very important work will be followed by at least one full paper and (one should hope) several papers. (All too often, however, this is not the case. All too often what is seen instead is a series of such short publications on a given topic, with few full papers.) It is essential that in your CV these important short STOP PRESS publications are clearly identified as such, and not confused with their humbler snippet cousins. (This can easily happen because of the structure of the refereed literature.) It is worth pointing out, however, that in some disciplines and sub-disciplines (e.g. biology and engineering), short papers and communications are not considered prestigious at all. In fact several biologist and engineer colleagues frown on our appreciation of short publications, noting that in their field "you either tell the whole story or you're not taken seriously."

Some journals publish exclusively letters or short contributions. Examples include *Applied Physics Letters* and *Physical Review Letters* for physics, and *Chem. Comm.*, *NanoLetters* and *Angewandte Chemie* for chemists. (Scientists from other fields will kindly excuse our lack of equivalent lists for their interests. This is another item to be attended to in a second edition.) Some other journals publish both regular papers and communications in the same volume, like the *Journal of the American Chemical Society* (better known as JACS), and *Physical Review A* through *E*, with their *Rapid Communications* sections.

Standing head and shoulders above and apart from these more specialized journals are *Nature* and *Science,* the two most prestigious scientific journals. These have a section devoted to Letters (Nature) and to Reports (Science), and a shorter section devoted to Articles, which tend to be longer contributions that report major advances in a given field (each issue only contains one or two of them, on average). They also have a section on very short communications, *Briefs* (Nature) and *Brevia* (Science) which are one page in length or less.

Generally speaking, in many (but not all) disciplines, Letter journals tend to be more selective, and therefore it is more difficult to publish in them. Precisely because it is more difficult, almost everybody would like to get published in a letter journal — the added difficulty and selectivity carry extra prestige and are often associated with a higher quality. The necessity of rapid publication (the original reason for founding these journals as fast-track vehicles) is now often slighted in the weighting of the likely impact and novelty of the publication. In fact, with appeals and corrections and the like, it is not rare to have some publications in letter journals actually take longer to see the light of day than the average time to publication in the associated regular journals.

A Letter journal generally offers the advantage that your submission is often (but not always) processed faster, and that your work, if published, because of the valued *imprimatur* of a highly selective journal, will be read more broadly (and hopefully more frequently cited). In the scientific arena, exposure of this kind is something everybody fights for. Being in the spotlight is almost everybody's dream. Peer recognition, as we keep repeating, largely determines your success.

On reflection, the tendency to write short contributions in certain disciplines is not at all surprising: most scientists, especially important and famous ones, tend to be incredibly busy, and are therefore unlikely to read long papers. Since famous scientists desperately want recognition from other famous scientists, they will invariably try to write short papers in the very best journals with the highest impact factors, so that a larger audience will read them; and so on.

Nowadays the most selective and prestigious sections in *Nature* and *Science* are called *Brief Communications* (Nature) and *Brevia* (Science) and they only take up about half a page and one journal page, respectively. The acceptance ratio for Nature's Brief Communications section is in fact roughly 5%, much lower than the Letters section.

Writing concisely and clearly is therefore an absolute must, particularly if you want to publish a letter. (Learning to write concisely and clearly is also useful when you apply for a fellowship or a grant, since most funding agencies provide strict guidelines about how many pages (or words) are available to write your proposal.)

It is a mark of respect for the community to write a long follow-up paper after you managed the arduous task of publishing a first letter. (This should be standard practice, but is not.) In this follow-up publication you will of course remind everyone that you just published a letter, and, more importantly, include all the experimental or theoretical details that simply could not fit into the letter format, but which are important if your work is to be thoroughly understood. This is particularly true if someone wants to reproduce your data or perform calculations based on your experimental results.

You may not want to go through the quasi-political hassle of writing a letter and arguing its way past the letter journals guard-dog referees, and you may therefore decide to write directly a long paper where the degree of hostility is lower.

Clearly, like the choice of journal in which to publish either the letter or the paper, the balance between the two is partly based on your own estimation of how important the work is (not all ducklings are unrecognized cygnets) and partly on your own taste for battle.[f] (Some cringe from battling referees, others relish it.) It is a good idea to evaluate your personal motives in making those choices. While you may feel detached about not pushing this particular piece of research to the Letter journal standard, you may be denying your graduate student a legitimate shot at a good start in their publishing career. Ethically speaking, given work of equal merit, one should probably push harder for the work in which a student or post-doc is the first author, since the immediate impact on their careers will be greater.

When writing a paper you should be *very* critical about your work, your approach, your results and the way you are presenting them. The best way to do this is to ask yourself, how would you rate this paper if you were to review it as an anonymous referee? Would it meet the standards of the journal where you wish to submit it? Would it have a fair chance of being accepted? Many small points of clarification in a paper are inserted to forestall a pointed question by a referee. (Answer the question before it is asked.) Again, think of this as a game of chess and do your best to be several steps ahead of your opponent(s) (in this case, the referees).

Of course, being objective about your own work is the tricky part here. Any scientist who has been even modestly successful will admit that their ability to write papers improved tremendously after the first few chores of refereeing are under their belt. After that it is much easier to see the flaws in your own work and in its presentation. For this reason you should be generous about acting as a referee; you will get as much benefit as the service you render. (Besides, it looks good on your CV.) Also, if your supervisor is doing a lot of refereeing, offer to help. Most will be grateful for the offer; but once you are experienced enough, it is best to make sure that it is *you* who sends the report in to the journal and thus gets added to their list of referees. (If you do a sufficiently good job

---

[f]We offer the Johnston Observation of Non-Reciprocity in Refereeing. "How is it that the journal editors send me such poor stuff to referee, while my submissions often fall into the hands of refereeing numbskulls who don't know excellent work when they see it?"

of refereeing you may eventually be asked to become an Associate Editor and this is a very useful addition to your CV.)

If you do this exercise of serious self-evaluation each time you write a paper, it will usually save you a lot of time later in avoiding delays inherent in making detailed revisions which would have to be checked again by the referee. A good paper has to be thought through exhaustively and should convince you completely when you submit it. A good way to do this is to write and rewrite the paper until you really cannot stand its sight any more.[g] At this point the best thing is to leave it for a week or two to "cool off" so you can regain your detachment before taking the next step. At that point, you are ready to submit, because it is unlikely that you can contribute to it any more. Another important piece of advice is to ask some colleagues (e.g. your mentor if you have one) to read it critically for you before submission. This "internal" review is important, and since it is informal and usually constructive, it is likely to save you a lot of time and frustration.

With junior colleagues as first authors, you should try to have them produce at least the first draft of the paper. After all they will have to learn eventually, so you are not doing them a favor by doing too much of the work. A strategy which often works is to sit down together and write the outline, and then send the student to write the paper from that. Of course it will not be as efficient as if you wrote it all yourself, but a very important part of the education to which the student (or a post-doc) is entitled is some training in paper writing.

As a general strategy, it is probably best to publish as many glittering Letters as you can, and, for the rest, it is better to publish a few good meaty papers rather than many average papers of modest length. (If people tend to say of your work, "Have you seen X's last paper on the "whatsit" effect?", you are publishing too many contributions so small that they risk being lost in the literature "noise." A good analogy is

---

[g]**Federico:** — This typically happens to me some time after the 30th draft, however I expect that each person will have a different tolerance threshold. Incidentally, when I submitted to *Science* in December 2001, together with my co-workers we went through approximately fifty drafts, and when we got the reports from the referees and the Editor, we were asked to rewrite the paper entirely!

maritime radar, where the echo from the waves is called "sea clutter." If the boats you try to see are too small, they will be lost in the "sea clutter". (The tendency we are advocating is that of the famous German mathematician Gauss, who had as his motto (on his seal): *Pauca sed matura* (Few, but ripe). If you are not as talented as the legendary Gauss, do not go to the extent he did. Many of his results were found in his drawers after his death, because he felt that he had not yet polished them well enough.)

Publishing papers of impressive weight will improve your signal-to-noise ratio, as well as your citation rate and your overall impact. (Of course it will reduce the raw number of publications and might bring harassment from the strict publication counters.) Psychologically it will also have a positive effect, since it will make you feel good about yourself and proud of your work. In the long run, you want to look proudly at your publication list, rather than view it as a collection of papers whose sole purpose was to advance your career. Graduate students often tend to fall into what we call the "short list" syndrome. It takes them a while to publish their papers, and they feel uneasy about having a short publication list.

Federico: — I used to feel like that when I was a student. This is understandable, since this list will be a determining factor in a student's ability to find a job after graduation. This is especially true if you want to stay into basic research. However, students tend to forget that in the longer run, the quality of their work — even their very early work — will largely determine their success in science. However if someone has a few lightweight publications at the start of their career, it will not hurt them in the long run, provided that the light-weight publications are phased out as the career gets up to cruising speed. (In any case funding agencies will often ask you to present only the last five or six years of your work. "What have you done for science lately?")

5.2.2 *The structure of an article/letter: Title, abstract, introduction, conclusions and references*

In terms of overall structure there is little difference between a Letter and full paper, except the length and the degree of detail, so the remarks here apply to both. The sequence given in the title above is important, because it gives the conditional browsing order in which a paper is usually scanned to be flagged for reading. A very busy scientist nowadays may not be able to go through the literature more than once or twice a month, and sometimes even less. (This is also very sad, but true.) To be flagged for reading the paper will have to elicit a "yes" from the reader at each browsing step or the browser will move on to the next paper.

In more detail, then, in browsing through journals, the reader will first skim through the titles. If the title attracts enough attention to warrant going further, the next move is to read the abstract, then the introduction, then the conclusions, and (perhaps) finally the references. (However, the references are often checked before the body of the paper to see if you have cited the reader's work, and to see if your knowledge of the literature is adequate, or perhaps even novel.) The body of the paper will often only be attacked if these preliminary indications are promising enough to make the reader think that it is worthwhile. Although you are not writing your papers exclusively to captivate and please super-busy scientists, if you do not pass these sequences of interest checkpoints, your paper will be read only by the small set of people who read everything on the topics they care about, including yours. You should want to do better than that.

The situation is like that of the store trying to lure a customer inside; the "browse" sequence being the name of the store and what it sells, any indication of a special sale, window displays, perhaps a display inside the store and finally the merchandise itself. In effect, the title should answer the implicit question in the browser's mind of each title "Why should I stop to look at this paper in more detail?"

The lesson from all this is that, when you submit a paper for publication, you should make sure that the title you choose is appropriate and captivating. It should be as short as you can make it, since longer titles are somewhat of a turn-off. (A superb title for review of some work on how frogs' eyes automatically track motion referred to a complex background was "What the Frog's Eye Tells the Frog's Brain." That is a title that is difficult to beat.) Remember that your title does not have to have too much detail, because that you can put into your abstract.

Of course, your abstract should also be short, clearly written, and should contain the main points of your paper. Your introduction (really the first paragraph if you can manage it) should place your work in its proper context, and give a broad view of why this field is important, and where it is leading.

Your conclusions are also important, because they may be the only thing most of your readers will remember. The conclusions may make the difference as to whether the paper is marked for a high-priority read, as something to come back to when there is more time, or to be copied into a running bibliography for the next paper the browser may be writing. Ideally, the concluding/summary section as well as the actual conclusions, should also point to new perspectives and directions of research. Finally, of course you should make sure that you are citing all the relevant literature, and if possible, even more. Remember, as we have said before, being generous in citing other people's work is very unlikely to do you any harm and can do much good.

Letters are so short that they require a lot of re-writing to get them right and yet keep them compact. With papers one can have dense patches for the expert and simple paragraphs to bring the less specialized reader up to speed on what is going on.

5.2.3 Dealing with referees

Having taken all the pains that you can, your *magnum opus* goes off to the selected journal and usually is returned with comments from the anonymous referees to whom you must reply (through the editor), and this is the principal topic of this subsection.

However, two other things may happen. Your work may be accepted exactly "as is" (a rare occurrence), in which case there is no more to be said. The editor may however declare without referee assistance that your submission is not suitable for the journal. This is most likely because the field that is being addressed is too far from the central theme of the journal, or (more rarely) because it is not up to the level that their referees need to be called to examine. In either case your dialogue is then directly with the editor whose name you know, rather than with anonymous referees, as transmitted through the editor. The dialogue is rather different and your part resembles that of an agent arguing for his client to get a publisher to look at a book or to obtain a part in a play or the like. You are in a difficult position with little negotiating power. Diplomacy, intelligence and perhaps cunning are needed, but it is difficult to give general advice.

Of course you might also run afoul of journal style rules, which most of us cravenly obey. In connection with journal rules (admittedly some time ago), an author was told (by a colleague) that a manuscript which he was about to send to *Physical Review Letters* would have to be modified because he was the sole author and used "we" throughout. Rather than switching to "I" (which was then not an allowable option) or changing it to the allowed impersonal passive voice (e.g. from "we have made mean-field calculations" to "mean-field calculations were made" etc.) which was judged too awkward given the use of typewriters rather than word processors, J.H. Hetherington chose to add his cat Willard as co-author F. D. (for Felix Domesticus) Willard. The full tale is told in More Random Walks in Science[1] on pp.110-111.

In another instance, the well-known physicist David Mermin recounted at length in *Physics Today* April pp. 46-53 (1981) his cunningly planned and successful campaign to get Physical Review Letters to accept "Boojum" from Lewis Carroll's *The Hunting of the Snark* as an internationally recognized term applied to a phenomenon in liquid helium-3 in phase A. (Amusing follow-ups of the kind frequently occurring in anything related to Lewis Carroll may be found in *Physics Today* September pp. 11-13 (1981), and March p. 96 (1982).)

Let us turn to the more usual case, which is the author-referee dialogue conducted through the editor. Clearly if only minor issues are involved the quickest way is to agree with the referee, make the changes and get on with your life. The difficulty comes when the disagreements are more serious.

Again the subject can be divided into two cases, responding in the first case to the referee who is in favor of publication, but wants specific changes with which you do not agree and in the second case to the referee who thinks the work is so flawed as to be not worth publishing.

For both these cases, the first piece of advice is to keep your temper. Do not rant either to the editor or to the referee; it makes about as much sense as shouting at Customs or Immigration officials, or the policeman who gives you a speeding ticket. While fair words may not succeed, foul words will most certainly fail. The second piece of advice is to try to put yourself in the referee's position and see through to the roots of the disagreement; this will be invaluable in putting your case in a conciliatory and civilized tone. The third piece of advice is to realize that the situation now resembles a jury trial, where you are the lawyer for the defense, the referee, the prosecutor and the editor is the judge/jury. The game can be won even if you cannot convince the referee to change the initial opinion, because the referee may lose credibility with the editor, as being unreasonably picky or shrill or even wrong. (This is more likely to be the case if there is more than one referee and the negative opinion is not in the majority.) All this is much easier to see and to do if you have done your share of refereeing and are thus used, so to speak, to "playing the game" the other way.

This possibility of loss of credibility of the referee during the dialogue is why it is very important to appear to be patient, reasonable and, yes, even sympathetic, with a tone that reflects more sorrow at a misunderstanding by the uninformed than anger at the insolent. (Remember that implying that the referee is not competent is an implicit reproach of the editor for not knowing of the incompetence or worse of the referee. The worst that you should imply is that the referee is perhaps a little out of his depth or obsessed on this particular point. Do not, for instance, wonder how this referee could have been picked to referee your work.) It also helps to take blame for not making the points sufficiently clear, even thanking the referee for bringing this defect of presentation to your attention, and so helping you to improve the paper.

In the case of disagreement on a point which is not a simple misunderstanding to be corrected, but strong disagreement of, say, interpretation (where difference is often possible), another tactic to consider is to include the referee's comment, but maintain your point with your reasons for inclining to your view rather than that of the referee. In effect, you are saying to the editor, "There are two possibilities here and we are presenting both and leaving it up to the reader." If the referee persists the editor may well decide for your ecumenism and against the narrowness of the referee.

If the referee is really negative, while you may try these milder tactics, there are other and sterner measures. If the referee's familiarity with the field seems shaky, you may undermine the credibility of the referee, perhaps by bringing other references and authorities that you hadn't included before, perhaps by phrases such as, "these objections have been dealt with elsewhere by etc." If the referee's opinion is too vague, and too sweeping ("lacking in originality" and the like) you can with justice complain of the difficulty of defending the work against such vague accusations without supporting detail.

If all these measures fail, remember that you can often demand the opinion of another referee. This should always be done in a tone that is slightly apologetic (for putting the editor to more trouble because of this stubborn referee) but firm.

All this is quite serious and stressful, so much so that a somewhat lighter look at the topic is worthwhile including for your amusement. The item is the well-known *A Note on the Game of Refereeing* by J.M. Chambers and Agnes M. Herzberg in Applied Statistics XVII n. 3 (1968), reprinted in *More Random Walks in Science*[1] pp. 8-13, and available (2005) in downloadable form at on the Web www.buzzle.com/chapters/science-and-technology_jokes-and-funnies.asp. Unfortunately the full text would take nearly five pages here, so all we can give is a sample or two to whet your appetite for the full text.

DIVERSION   Excerpts from *A Note on the Game of Refereeing*
   ... It is agreed that the author's objective is to have his paper published, and that extra points accrue for the publication of a particularly worthless submission. ... Likewise the referee's minimal objective is to have the paper refused and extra credit is obtained if the paper was a major contribution to the field. ...

After the opening, it is worth sampling more.

DIVERSION More excerpts from *A Note on the Game of Refereeing*

Author tactic A5: A5. Flattery-may-get-you-somewhere tactic In the revision of the paper the author thanks the referee for his "helpful comments" etc. This is very often employed against tactic R5 (deliberate misunderstanding of something which is correct) by saying something to the effect that he (the author) "agrees that he was not clear in the earlier version of the paper."

A7. Precedent tactic. Reference is made to a paper which although of very low quality was recently published in the same journal. The author implies that his work cannot be of lower quality than the previous paper. The danger, however, is that the editor may be only too aware that he should have rejected that paper and will act accordingly.

Referee tactic R2. Wrong-level tactic. No matter what degree of rigour the author uses, the referee replies by saying that it is not the correct one. For example, "The author has stressed rigour to the detriment of clarity," "The author's colloquial style is insufficiently rigorous," "The author unfortunately tries to combine rigour with a colloquial style to the detriment of both."

CONCLUSION ... It must be acknowledged that the entire practice of referee-man-ship has declined in recent years. With the publication of more and more journals, and the issuing of present journals more frequently, the pressure for papers to fill them restricts the referee from rejecting as many acceptable papers as hitherto. ... However, the most insidious cause of this decline is the loss of the true savage refereeing spirit among the modern generation of players. We fear that too many participants have taken to heart the old adage, **"Referee as you would others referee when you are writing."**

5.3 Ph.D. theses

Most people tend to consider having to write a book-length thesis as a major obstacle to their progress imposed by an unfeeling university. The thesis needs to be dealt with to get their degree and many would gladly trade it for a thesis composed of stitching the relevant papers together with a bit of integrating text. (But see below in Section 5.3.2 for our contrary opinion.) The student might also say, "If the stuff is good enough to publish shouldn't that be enough?" The short answer is "no." The student is supposed to have reached the point where they could do autonomous research; it is the student that must be examined. Hence the thesis and ritual examination and presentation are necessary.

Insofar as a thesis demonstrates anything, it is supposed to demonstrate to the thesis examiners that you actually understand what you did, appreciate the context and did not behave merely as a super-technician following your thesis advisor's directives to the letter with no thought of your own. (The flaw in this reasoning is of course that, in that case, the thesis advisor could micro-direct the writing of the thesis just as well.)

Hence the importance of the questions associated with a thesis defense, when the candidate is supposed to respond without assistance from his thesis advisor. (Since in fact theses which survive to examination are hardly ever subject to more than extensive corrections at worst, this aspect of a thesis examination is usually more formality than fact, more ritualistic than rigorous.)

In a well-run doctoral system, if the advisor has missed a significant difficulty, the humane solution is to postpone the thesis defence and fix the problem(s). Thus by the time the thesis is formally defended, the serious difficulties should be all ironed out.

If you write a good thesis however, you will be performing two and perhaps three useful tasks. It is your first (and in some cases only) chance to write a *comprehensive* text on work carried out over a period of several years in useful detail.

First of all, the work in organizing all your efforts into a thesis which is far longer than any paper and which thus allows for a much more self-sufficient treatment, will stand you in good stead when you have to write a comprehensive report later on in your career at the end of a major project.

Second, most (but not all) theses follow previous detailed work by other candidates in the same group, usually with the same thesis advisor, as you used the previous theses as a detailed guide to the development of apparatus and procedures and perhaps computer programs, now is the time to contribute your share to your advisor's group and to future students.

Third, should the work prove to be so seminal to the scientific community that others will wish to follow it in detail, then the detailed treatment in the thesis will make plain to interested readers what is only sketched in published papers.

As just indicated, writing a thesis is quite different from a scientific article, and not just because of its length. A thesis in fact is a much more comprehensive body of work than most papers. In your thesis, you should describe carefully and thoroughly all the work you carried out as a student in Prof. *Seldom Available*'s laboratory. This is a good place to include all sorts of experimental or theoretical details and approaches that for some reason or another cannot find their way into your published papers. It is also a place where you can discuss things which did not work and why, details usually squeezed out of papers by the editors' pressure to compress manuscripts. Other valuable information may include a new data analysis method that you developed, or an improvement of the experimental technique you have used.

If your contributions are very important, they can (hopefully) be published as regular papers in peer-reviewed journals. If on the other hand you developed something new but not terribly innovative, the right place to record it is your thesis. It may prove of value to other students and scientists later on, so it is still worthwhile to record it in some detail.

When you started your graduate studies in Prof. S. Available's lab, was there a thesis from a previous student that helped you get started, perhaps with descriptions of complicated procedures? If not, would you have benefited from having this type of information at hand? More than anything else, science builds on to previous knowledge,[h] and in a scientific team, well written graduate theses can be extremely useful in keeping the continuity of the laboratory. Taking this to heart, you should try to make your thesis of somewhat of a do-it-yourself manual for your successors. If you write a good thesis, including a great number of details and a thorough description of the procedures you used, your work will be useful not only for you but also for the next student who takes your place in your advisor's laboratory, and continues your work from where you left it.

To sum up, your thesis is your first chance to learn how to write a comprehensive body of work, describing in detail all you have done in a period of about three (or more) years. Sometimes, a good thesis can actually be transformed (with quite some work) into a review article. Thus, you should definitely take advantage of this chance, and see it as an opportunity to learn rather than a burden. In our opinion, it is an important part of your scientific training, which eventually will earn you your doctoral degree.

If you want other scientists in your field to know you, and to appreciate your contributions fully, you may want to circulate your Ph.D. thesis among them. As a first-order approximation, your published papers will be more in demand (assuming that they are good of course). However, as we discussed above, your thesis will probably contain the detailed procedures you used and all sorts of information that will give much better clues about your maturity as a scientist, especially if someone is trying to decide whether to employ you as a post-doctoral fellow.

5.3.1 *Language of the Ph.D. thesis — English!*

With few exceptions, a Ph.D. thesis normally does not have a huge readership. In general it can only be understood by experts in the field, which probably limits the total audience to about 100 people *worldwide*. Thus, if you do not write your thesis in English, a language understood

---

[h]Remember Isaac Newton's quote: "If I have seen further, it is by standing on the shoulders of giants." (Also on the British two-pound coin.)

by any ambitious scientist, this will limit your audience even further. Although in the short term it may be seem more useful to write it in a language other than English, either because it is the language that the local students speak, or because it is your mother tongue or because you have mastered that language better than English, in the long term the opposite will likely be true and the thesis will remain of strictly local use.

Writing your thesis in English will certainly help you improve your writing skills, even if English is your native language; and this in turn will help you in most of your future scientific endeavors. Considering that English is widely accepted as the international language of communication for the Natural Sciences and Engineering, mastering this language both orally and in written form will become an asset, and in the long term it will help you succeed as a scientist. A thesis in the local language may well have a local use, but few others will be able to use it.

A final point is the following. Since so many students come from other countries, and since the language of science is English, if the next student comes from, say, India or Brazil or China, the thesis will be of immediate use to that student in a way that will not apply for a thesis written in the local language. (On the other hand, it could be argued that the effort to read a thesis in the local language may help the student to acquire competence in reading the local written language.)

Although the circumstances no longer apply, the general principles in the following anecdotes from Federico on language and science may be useful.

Federico: — In connection with this question of language, my grandfather was also a scientist (a physicist, for a change!), and he worked in Italy between 1930 and 1964, approximately. At the time, there still was no unifying language for Science, and he had to learn no less than English, French and German (and even some Russian) so that he could read the relevant papers published in foreign journals. At that time, scientists like Einstein, Heisenberg and Schrödinger published in German, whereas De Broglie published in French, Fermi in Italian and so on. Since he had to spend so much time learning other languages, clearly this slowed down his scientific progress. So you should not be surprised by my firm belief that it is a tremendous advantage to have a unifying language for the natural sciences and engineering.

Example of T.'s thesis

I was once visiting a group of colleagues in France, and I happened to be in the office of an Italian scientist, T., who has a permanent position there. He had previously done his Ph.D. at a prestigious Institute in Germany, where he had worked with one of the fathers of Surface Science. I knew his work fairly well, and had read enthusiastically his papers published in the very best journals — *Science, Nature* and *Physical Review Letters.* Since I was curious about certain specific aspects and details of his work, which had not appeared in his published papers, I thought I might find them in his Ph.D. thesis, which could become a useful reference for myself and other colleagues. Upon request, he proudly produced a copy of it, telling me that it had been a great achievement for him to be able to write it in German. As you can imagine, I was profoundly disappointed. Although his thesis was a small work of scientific art, I doubt that anyone else ever read it besides his advisors and his opponents. He still offered me the copy, and I politely declined.

Two European examples: Italy and Denmark

When I was a student in *Italy,* the rules for submitting a Ph.D. thesis had just changed (thankfully!). The novelty was that students could decide which language to write their thesis in (the choices were either Italian or English for the natural sciences and engineering).

This means that I did not have to apply to a committee, asking permission to write *my* thesis in English. *I just did it.* And quite honestly, in my opinion this is the best possible approach. A graduate student is supposedly a grown-up, mature person, and since the language in which his/her thesis is written will mainly have an impact on his/her life and career, it should be entirely their decision. Asking permission to a committee, on the other hand, implies that this permission may actually be denied, which I find unacceptable. Why should a committee be allowed to decide on your behalf something that will impact *your* career?

On the other hand in *Denmark*, where I worked as a post-doctoral fellow for one year and a half before moving to Canada, there is no choice: everybody has to write the thesis in English. Denmark is a very small country with an outstanding scientific tradition; most of the students who were working in my same group had already developed excellent writing skills in English, which helped them write compelling papers that were published in the very best journals.

5.3.2 *Thèse par articles*

In some universities one can submit a thesis consisting essentially of published papers. This is called a *"thèse par articles"* (i.e., "thesis from publications," the term we use from here onwards). To us this thesis from publications has the appearance of a thesis choice made by some particularly lazy person. A thesis by publications consists of the publications published by the student throughout his/her graduate work, together with introductory material and some conclusions, stapled together in just one file.

Our advice is to discard completely this possibility, and to opt for a full thesis instead, on the grounds that a thesis by publications is not a *real* thesis and is essentially very little more (just the "glue" text that holds it together) than the sum of its publications. It is true that the classic thesis requires much more work than the other. However, with the classic thesis you are investing your time on a useful endeavor, instead of wasting time (admittedly much less time) on something that nobody will ever read or request (except in error for a real thesis). In fact when we write something, our aim is to provide some useful information to a target readership. If on the other hand we should believe that nobody is ever likely to read what we are writing, we would be better off doing something else entirely.

There is, however, one special situation where a thesis from publications can be a useful compromise, and that is where the student's grasp of the local language is not good but that the option of writing a thesis in English is not available. In that single case a thesis from publications minimizes the amount of the local language that must be used. However it is clear that, while easier for the candidate, as remarked above, the local utility of the thesis over the mere sum of the published papers is likely to be negligible. This is therefore an inferior compromise solution, but obviously better than nothing.

5.3.3   Structure of the Ph.D. thesis

In the introduction, you should clearly state why you embarked on this project, and what the challenges were that had to be faced when you first started. A good introduction can become excellent reference material for you and your peers.

In the body of the thesis, you should report the methodologies you have used, the issues and problems that you were confronted with, and how you set out to solve them. You should include all the details that you think are important for someone to understand your work, to reproduce your data, and to continue on from where you left the work. If you developed a new technique (e.g. fabrication, processing, characterization, data analysis, computational algorithm) this is your chance to describe it in detail, since very often there is not enough space for such a thorough description in journal articles. It should be noted that there are also cultural differences between Europe and North America in how extensive the candidate should make the review of the field. (Europeans are required to make quite extensive reviews of previous work, presumably to demonstrate that they understand it well.) It would seem best to abide by the local custom. Since there is usually no upper limit to the number of pages you can write, this is your opportunity to write extensively and exhaustively. Somewhere between 100 and 200 pages appears to be the norm.

Finally, in the conclusions you should clearly identify your contribution to the field, and outline what are the future perspectives and challenges. If you manage to do all this, you will have written a good thesis, and although it may have a more limited circulation than the papers you published in peer-reviewed journals, it may actually become a useful read.

5.4 Curriculum Vitae (CVs)

As remarked at the beginning of this chapter, the CV is the traditional way to communicate your worth and provide the links to your work for such vital aspects as employment, fellowships, scholarships, prizes and the like. In general, apart from a limited number of copies of published papers and the actual text of the proposal/application, the full background is encapsulated in the CV that accompanies them.

There is no reason, however, to use one invariant form for the CV and it is a good idea to "prune the tree" of your basic source CV "tree" with its very complete "trunk" and thick "branches" to tailor it for the job it is to do. (It is much easier to select than have to chase the data afterwards for details.)

For some applications the judicious selection of your recent work is what is of interest, while for others completeness is necessary. Creating and maintaining the full CV tree is a necessary and ongoing chore, while the tailoring of the CV for particular cases is an episodic process, according to requirements. Let us first discuss the basic CV tree.

5.4.1 The CV tree, offshoot CVs and CV components

The CV is often referred to as if it was a single object, growing by accumulating, as in "I'll have to update my CV" or "That wouldn't look good on my CV." (By the way, in North America CV is often synonymous with *resumé*.) This is not true. There is indeed a central CV complex here called the *CV Tree*, with several (possibly many) *CV components*, including but not limited to the following: *Education, Employment, Teaching, Awards, Refereed Publications, Refereed Contributions, Invited Presentations, Books and Book Chapters, Seminars, Ongoing Projects, Future Plans, Collaborations, Teaching Experience, Current Students, Former Students, Funding* and whatever else might be relevant. If you are farsighted, you will continually update all the components of what comprises the *CV Tree* as changes occur. Associated with this *CV Tree* are various subsidiary or special-purpose CVs or *Offshoot* CVs created for special purposes, and these are as varied as the uses to which you might put your CV.

The point here is that for many uses only a fraction of the CV components are needed. Also in many cases not all of a particular CV component is needed, it frequently being the case that one is restricted to data such as publications or funding applications only for the last few years. Often you find you need to put a CV together in a short time, and it is much easier to do this if the components on the main CV Tree are updated regularly. Sometimes one only has to update a special-purpose Offshoot CV from a previous application without having to go back to the original CV Tree source. Let us take these CV components one by one, but before that, one question should be settled and that is the order in which the data is presented in each CV component.

Should the elements in each component be given in chronological order or in reverse chronological order? The safest way is to be redundant and to choose both and update both on a regular basis. If you are asked to provide a CV for a lifetime achievement award then the chronological ward seems only natural. However if what is of interest is only the last few years, as is often the case, then the reverse chronological order has much to recommend it, at least for publication and funding. For instance, take refereed publications (including refereed conference proceedings). One should of course maintain the chronological list (this sometimes gets complicated because papers may not appear in the order that they were submitted) with strict and immutable numbering. This has the important advantage that these numbers can be used as reference or citation numbers forever, and these permanent numbers can be used in the body of the CV when discussing accomplishments or future plans for proposals and the like. On the other hand for cases when only recent work is to be discussed, presenting the data with the most recent and most relevant first has much to recommend it. (Of course the publication numbers are prominently positioned at the left, probably in **boldface** if permitted, for ease of reference.) You add the new work at the top and drop off old work at the end. The numbers also serve to remind the reader of how many total publications you have. In the same way, a prospective employer only cares about the last one or two employers, not what you did twenty years ago, and the funding agencies have the same interest in the recent past and not the distant past. While the safest course is to maintain both orderings for all components, but to use mostly reverse ordering in the *Offshoot* **CV**s to keep the presentations manageably short.

5.4.2 CV components

Education, Employment, Teaching, Awards, Refereed Publications, Refereed Contributions, Invited Presentations, Books and Book Chapters, Seminars, Ongoing Projects, Future Plans, Collaborations, Teaching Experience, Current Students and Post-Docs, Former Students and Post-Docs, Funding

Education This is pretty standard, but many people omit their thesis title and thesis advisor. They should be recorded on the **CV Tree** at least so that they can easily be added for a particular case.

Employment This is again standard, but still, if there are people with whom or for whom you worked, one should note the names for possible inclusion in a particular case.

Teaching While non-academic employers are not interested, universities naturally are. Again if there is someone who can usefully comment on your teaching experience, they should be included here, in case they are needed in the future.

Awards Should be indicated for all employment opportunities

Refereed Publications As discussed above, this is a key element in all CVs. The only questions in a given case is whether to give all or just the recent work, and whether one uses chronological order or reverse chronological order, whether to give paper titles, and whether to give finishing page numbers and how to order the placement of the components..

Refereed Conference Contributions Since they are refereed, they should be in the publication list with their individual numbers.

Invited Presentations Like the refereed Conference Contributions, they form part of the list of refereed publications with their individual numbers.

Books and Book Chapters Although implicitly refereed for the publisher, these are not considered original refereed publications and should not be numbered with them.

Seminars After a while the number of seminars which in content duplicate the publications becomes irksomely large. Probably one should put all the seminars in for the early years, and keep only those from the last few years (say, five or ten) after that. A more magisterial approach is to say something like "Each published paper has, on the average, been the subject of about *N* presentations at conferences and seminars." On the other hand you would like to note the seminars before particularly august assemblages and in prestigious institutions.

Ongoing Projects For something like a possible employment or cross-over appointment dossier, an outline of your ongoing projects is indispensable.

Future Plans For something like a possible employment or cross-over appointment dossier, an outline of your research plans is indispensable.

Collaborations For many purposes a summary of your ongoing collaborations (including institutions and researchers) helps in defining and clarifying your research activities and shows how well you are regarded by other institutions and researchers. (Of course this may well be evident if one looks carefully at the list of authors in your publications, but the aim is not to force the readers to have to dig this out by themselves.)

Teaching Experience This is indispensable for academic employment if you have not done very much of this, being preoccupied with research. Universities will always want to be reassured that you can really contribute to their teaching.

Current Students and Post-Docs This helps to indicate the size of your current empire. It is probably useful to indicate where the students were before and yet more importantly, where they end up after they leave.

Former Students and Post-Docs Again both future students and post-docs might like to consult your former people. However it is not easy to keep up with the changing addresses of former students and post-docs after they have left.

Funding Funding agencies often want to be reassured that you are not "double dipping," by getting money from two sources for the same work and using the extra money to do something else. Often a current summary of totals is enough, but this is just about as hard to keep up to date.

5.4.3   *Tailoring your CV to the purpose at hand*

"*Know thy neighbor*" Here is some simple advice on how to write your CV for a particular purpose, using as a resource all the CV components that you will be keeping up to date, and keeping in mind that it should be written as a function of the target audience you would like to impress. That is why it is essential that you "know your neighbor" well enough to fashion an appropriate CV for the purpose.

Writing a CV would be relatively easy if all that was required was bald listing of your assets and career to date and if the same CV would serve all purposes. In fact, writing a CV which is *well adapted* to the purpose at hand requires some thought, but the reward for this effort can be extremely important. In particular, you must be aware that you should write differently, depending on the intended recipient. For example, a CV intended to land you an interview for a faculty position should not emphasize the same achievements as a CV intended to land you a position in industry.

There are cultural differences to consider as well. As mentioned above, in a CV intended for a North American University the text should be written a lot more "aggressively" than for a CV written for a European University. In Europe it would seem that modesty is a quality that is still appreciated. If your CV indirectly boasts that you are a genius, and your reference letters support this claim, your European peers will probably wonder why you have not been invited to Stockholm yet, and perhaps frown upon you. On the other hand, if you are too modest in your CV when you send it over in North America, it will be trashed immediately, because people will think you are simply not good or ambitious enough. Therefore even cultural differences can be very important when looking for a job. Again, it is important to be aware of them and whenever possible, to use them to your advantage.

Your aim is to place yourself as best you can on the job market or in the list of applicants for a fellowship or award. Through your CV and perhaps an interview you are literally trying to *sell* yourself to a prospective employer or fellowship/award committee. You have to be convincing, because the people to whom you are applying have all your competitors to choose from, and they do not want to make a mistake in their choice. Remember, in seeking employment, or a fellowship, or an award, it is not good enough for you to do well; you actually have to do better than everybody else! Thus, it is potentially much tougher than just passing an exam or even getting a good grade, which were your (less ambitious) aims while in school.

Your CV, or *résumé*, should begin by describing in detail what you have done, but it should also give a clear idea of where you want to go from there. If possible, try to build it up so that it shows what kind of *vision* you have for your future. Interviewers like applicants who look ahead, instead of focusing on the past. In this sense, having a glorious past is generally not enough to land you a job: in your CV and during your interview you will have to show how you intend to build on the past. Your vision does not need to be correct or even accurate, but it is very important to show that you have one, i.e. that no matter how young you are, you actually take time to look into the future and plan ahead.

The main difficulty in writing a good CV is that you have to be concise and complete at the same time. You want to tell your prospective employer about all the important stuff that you have been doing, and outline your future perspectives, but at the same time you should do it in a few pages at most (excluding your publication list, which, by contrast, will hopefully fill up many pages). Unless every single line in your CV describes a breakthrough achievement, after a few pages you will lose your audience completely, either out of sheer boredom or lack of time.

When a University advertises a new faculty position for example, it is not uncommon that the department receives more than 100 applications.

Usually each application will be composed of a cover letter, a CV, a statement of research interests, a statement of teaching philosophy, and several (typically three or four) letters of reference (usually sent separately). All in all you can expect a minimum of 10 pages to read per applicant. (This is really a minimum; we were recently part of a search committee and would say the average number of pages per applicant was about 15, with peaks of 40 pages in some particularly unfortunate cases.)

You can imagine that the selection committee will have a hard time looking through all the applications in detail, especially if they are long rather than compact.

Thus if you manage to say all you need to say, and be concise and synthetic at the same time, your CV will definitely stand out, and this will increase your chances of getting an interview (as long as there is enough substance in your past activities, of course).

On the other hand, if you write too much, unless everything you say is really important, the members of the search committee may get bored and move on to the next application in the pile. This is again related to the concept of increasing your signal to noise ratio. If you do it well, you will have a great advantage over your competitors.

Incidentally, when you submit a grant proposal, the funding agency you are requesting support from will generally require that you attach your CV to it, and they will provide strict guidelines about the format (margins, font size, etc.) and overall space you should use. Once again, you are expected to write exhaustively about yourself, but to be concise at the same time. To obtain a somewhat different perspective, read the section on scientific writing. There we describe the difference between writing a letter and a regular article. Writing accurately, concisely and exhaustively is a very useful, perhaps necessary (but not sufficient), skill to become a successful scientist.

5.5 Oral presentation and organization

Much has been written[6] on effective presentations in front of an audience with images on a screen with the presenter controlling the timing and the sequence of the images. Nonetheless there are a number of points which do not seem to be given enough emphasis when discussing scientific presentations before audiences of significant size, and these points are what we discuss next.

The first important thing in an oral presentation is to be very sure of the allotted time and never to exceed it. (It is in any case most discourteous to the other speakers (in implying that your work is much more valuable than theirs) and to the organizers to go over time.) To begin with, you will almost always have to respect severe time constraints when you *perform* at the real conference. (Small-scale working groups and workshops are often much more relaxed with respect to time.) In fact in most meetings nowadays oral presentations are allotted between 10 and 15 minutes, including questions and discussion. We have all seen talks interrupted well before their intended end by zealous chairmen who were trying to respect the schedule. Some chairmen do it regretfully, others are most unceremonious.[i] You certainly do not want that to happen to you, both because it is embarrassing and because you would not be able to tell your whole story. (A book without its last few chapters does a bad job of getting the message across.)

To be able to deliver your talk in the allotted time, it is essential to practice your talk — or your poster presentation — at least once, possibly more, with a local audience which is *friendly*, but one charged with the task of looking for problems in the presentation, including time. If they are nice to you and grill you hard enough, there is a good chance that you will feel comfortable giving your talk in front of an arbitrary audience. This confidence will greatly increase the likelihood of a good performance. Also, this initial trial may even expose the weaknesses and occasionally the pitfalls in your work and how you present it (confusing images etc.), so it may help you to make significant improvements in the whole presentation.

Most of the advice on giving talks[6] focuses on what you should **NOT** do in a presentation. You should be clearly aware of what the most common pitfalls are. (There is some interesting, even funny literature on this subject, as, for example, *"How to give a truly terrible talk"* and *"Fifteen ways to get your audience to leave you,"* both of which can be found fairly easily by browsing the internet, i.e., *Googling* in practice.)

You should never overestimate your audience. In a sense you want to take the audience from a place in which they are comfortable to your space probably at supersonic velocity but without their realizing that they've been through the sound barrier. Like most people, although scientists like to learn new things, they do not like feeling ignorant or

---

[i]Being a chairman at a conference is considered by many to be a prestigious assignment, but it is also quite onerous and tedious. You have to sit through the whole session (as opposed to roaming through other sessions, networking in the corridor or even going to the bathroom), and listen carefully so that you can ask questions in case nobody else does. You also have to keep the schedule (which is arguably your most important task) and moderate the discussion, especially if some controversy arises.

stupid (well, after all, who does?). Therefore it is wise to give a broad but compact introduction, especially when giving a full seminar, describing in appropriate detail the state of the art in the field, and where your work comes in. You should explain clearly why this field is promising, perhaps what prompted you to pursue this topic, and what type of contribution you are giving. To clarify what is new in your work, you have to begin by placing it in the proper context.

In giving your presentation, you should be telling (in some sense, *selling*) a story. This means that your talk should have a clear beginning (in the form of an introduction), a middle section, and an end (in the form of conclusions and hopefully also perspectives for future work).

It is often hard to fit all your material, and to tell a good story, in the short time allotted. (A typical time slot is 10 to 15 minutes or so for an oral presentation, especially at big conferences like the APS, MRS, AVS, ACS, EPS, ECOSS etc.) Nevertheless, the rules are the same for everyone, so you should adhere to them and if possible, take advantage of them. In this sense, particularly because of this very stringent time constraint, our best advice is to try to present just *one* new idea or result.[j] If your audience goes home with a decent understanding of this one concept, you can consider it a very good accomplishment and your participation in the conference will have been worth its while.

Since time is short, you should make sure you are conveying only the really important concepts, and that you are not providing too many irrelevant details that would clutter your presentation. In fact, if your talk is appreciated, someone from the audience may come up to you *later* to ask about the details. (One easy solution is to provide a reference to a source for details, such as your e-mail address or even a presentation on your Web page.) After all, when you are finished, you definitely want the audience to remember the key points of your work, and not the petty details. If, on the other hand, you submerge your audience with an ocean of technicalities, it is unlikely that anyone will look you up later to find out more about your work.

---

[j]This is also true in relation to writing articles. If you include too much information, your paper will quickly become confusing and difficult to read.

You should use simply presented graphs or images as much as possible.

Perhaps the worst offenders are theoreticians who often tend to present too many equations. These quickly become a distraction and tend to attract time-wasting remarks on their nature. The best theoretical talks we have ever heard showed little or no equations at all, and focused almost exclusively on *concepts*. It is something difficult to do when you are young and inexperienced, however this should be your aim. The sooner you learn this lesson, the better. (Also you will be implicitly displaying your mastery of the field by showing that you don't feel the need to have the equations in front of you in case you forget them.) (Of course, if the basis of your talk is a well-known equation with a modification, you are allowed an equation or two to make this clear, but control the urge to go further, except when your audience are in your sub-specialty.)

Experimentalists sometimes sin in a similar manner by showing far too much detail in the sections on experimental arrangements and procedures. (A neat trick in computer presentations that can be used to control the complications is to use the Power Point facility that allows you to bring objects to the screen, to show the block diagram, zoom in on particular blocks for some necessary detail, and control the temptation that arises when the whole detailed diagram is up at the outset whereupon many in the audience will be trying to understand something that is not what you are talking about. Of course this strategy can also be used by a theorist for equations.)

Be careful of color. Many men are color blind and may confuse colors you think are quite distinct. Often the lazy option of colored graphs will give some colors (such as yellow) which are hard to see particularly if the lines are thin. Complicated background color schemes can confuse the perception of foreground objects. These are all things to check in your rehearsal presentation(s).

Do not read word for word from your slides, except for a short section where you are trying to emphasize something particularly important. (Remember how irritating it can be as a spectator, when the speaker reads from something which you have already read.) Most of the time, simply commenting on certain aspects of your viewgraph is enough to give an idea of what you mean, since your audience is presumably able to read. While it is a very good idea to prepare a guided discourse, you should not read from your notes! You are not in high school any more. You must look and sound professional.

If using actual physical transparencies (rather than using computer projection), it is often convenient to separate the transparencies by black and white paper copies, to remind you of the contents of the next transparency. These paper interleaves are also ideal for scribbling notes to yourself reminding you in writing about something that you want to mention, but which you did not put on the transparencies.

On the other hand, if using something like Microsoft Power Point, the 6-frame paper handout summaries of your talk remind you of the conceptual framework of your talk and allow for the odd note to yourself. These handouts can be cut into the individual slides which are a very convenient size for hand-sorting into a different order as you are organizing your talk.

If it is possible, and if it makes sense, you should use any help you can from modern technology. Power Point is used more and more frequently these days. It enables you to couple some special effects to the actual contents of your talk. Of course you should not exaggerate — your object is to sell your science, not to distract from it.

It is always wise to bring with you conventional transparencies as a form of backup in case Power Point or the projector system fails. (Of course that version of the talk would not be able to display the clever dynamic effects available in Power Point, so you should keep that in mind when making your emergency conventional transparencies.) It happens rarely, but if it were to happen to you … .

If you do not feel comfortable with having to give a talk in English, especially if it is not your mother tongue, you should take care to rehearse enough times so that you build up the necessary confidence. We say this in the hope of not having to sit through more talks during which the speaker is actually *reading* from a script …! (But then again, people who "read" from memory also tend to be quite boring, even if their English is good.)

If you become a good speaker, and do good science, you will be invited to talk many times. Besides the positive effect this will have on your ego, it will also help you further your career.

We hope that the foregoing will be a useful addition to your stock of knowledge on presentations.

Another and striking point of view is that expressed by David Mermin's *alter ego* Bill Mozart in a Reference Frame piece by Mermin in the *Physics Today* issue of November (1992) on pp. 9, 11, commenting to some extent on Garland's well-known remarks[6] on talks.

Among other thought-provoking remarks there was one which was particularly striking. "Give yourself a week. If you still can find no reason why anyone not directly involved in the work should find it tediously obscure, then you should find something else to talk about. Indeed you might seriously consider finding another area of research." (Although this little fragment had been planned as a *DIVERSION* here, it seemed that it might be too sensible to characterize it as such.)

5.6   Poster organization and presentation

While much has been said about oral presentations, not a lot is available in print on *posters*. On the Web however there is a fair amount.

An appealing source is one *Advice on designing scientific posters* by Colin Purrington, (Department of Biology, Swarthmore College, Pennsylvania) evidently designed to help poster presentations for scientists (biologists) from Swarthmore: www.swarthmore.edu/NatSci/cpurrin1/posteradvice.htm. Among other excellent features there one can find references[7] to two books (only one explicitly on posters) and five papers dealing with posters.

The particular strategies we recommend for the presentation and use of posters will now be discussed in some detail.

A poster should not be constructed by going through a talk with something like thirty images and then laying these out (one hopes in numerical order) on a poster surface in a left-to right rows, piled top-to-bottom like a television raster. This ignores the fact that a poster session is really more like a bazaar with many competing vendors. Unlike a bazaar, however, (but in the same vein as the two-public model for your targets for texts) there are two different classes of poster (bazaar) customers. They are, roughly, the professionals (those who know quite a lot already about the topic and are interested in the important and variant details) and the amateurs (who know next to nothing). Also poster sessions can be crowded (at least locally), and this means that the lower part of a poster space may well be blocked by people and can only be seen by those in the front row, right next to the poster and presumably the most interested. This suggests the following strategy, here dubbed the *Stalactite Strategy*. (The specific implementation below is based on the use of basic building blocks in the form of the usual 8½ by 11 inch (or the European A4 format) paper images in landscape orientation — better for large print — as building blocks, easily obtained from, say, Power Point.)

The strategy is similar to that of a shop in a street. One puts the summary and spectacular images in the shop window where they can be easily seen by passers-by.

For a poster this means put this key stuff, just above head height, so passers-by can see it easily (the "shop window"). The top-line story runs from left to right and summarizes what you want to say in something like six simple landscape images. The sign-up sheet for requests and envelope for business cards should be in the farthest right column, three down from the top. Each column (four or (perhaps) five images deep) goes into more intricate detail as you go down to the bottom. Altogether this is the *stalactite* mode of presentation (remembering that stalactites are the ones that hang *down* from the cave ceiling). With a few arrows and a bit of extra text one has a poster which works in a crowd and can be understood even in the absence of the presenter (the reason for the arrows). (When filling requests for an e-mail version, the images are rearranged for a serial presentation as given by the image numbers which Power Point readily provides and which you should always use and display.)

DIVERSION   Once again Stanley Harris has a relevant cartoon. Here there is no caption but a sign (inside a large cave with many people) which reads as follows: *"STALACTITES grow from the CEILING, STALAGMITES grow from FLOOR — PLEASE DO NOT ASK THE GUIDES WHICH IS WHICH."*

(By the way, it is easy to remember (but is little recognized as a mnemonic) that the vertical part of the "t" in "stalactite" looks as if it is hanging down from a roof like a stalactite, and vice versa for the "m" in stalagmite.)

Of course, when you prepare your poster, more or less in the same way as you do when you prepare a talk or write a paper, you should make sure that you organize it in such a way that you can tell a simple, effective story when somebody shows up to hear about it. (Surprisingly, some poster presenters do not have anything prepared beforehand about their poster. This is almost insulting to the clients, somewhat like having ignorant sales clerks in your shop. Not good for sales.) A lot of people will, of course, just glance at your work and then pass on to the next poster. However some, hooked, as it were, by the top line of images, may stop and ask questions, and they are certainly entitled to hear a coherent story. In this sense, presenting a poster is very similar to presenting orally. One difference is that again you should have prepared two levels of talks, one for the experts who want the newest details, methodology and the like, and the other for the tourists who are prepared to be entertained, but not too profoundly.

To make sure that the people who come to see your poster do not forget about you and your work, in addition to the sign-up sheet for requests, you should have with you some reprints (mostly for the experts) of the work you are describing in the poster, together with a considerable number of business cards with your e-mail address on them (among other things). (Business cards are a "must" at any conference and even more for a job interview.) If your visitors like your work they may actually end up reading your papers on the subject and either offering to collaborate or at least citing your results in their own work.

Chapter 6

Cautionary Tales

Sections of this Chapter 6

Federico: — In this Chapter, while Tudor takes a holiday, I relate a few episodes on M.Sc., Ph.D., postdoctoral and job Interview experiences from friends and acquaintances.

6.1   My summer work in Japan (Federico)

During my work as a Ph.D. student (Nov. 1997 - Oct. 2000) I went twice to Japan as a summer student (1999 and 2000). This was a useful and challenging experience. It gave me the opportunity to work with other scientists and learn new approaches, on one hand, and to travel to an exotic country like Japan — which was very interesting from a cultural point of view — on the other hand.

Not everything went smoothly however. (Here I will only dwell on the professional issues, even though the cultural and personal ones were perhaps more interesting if not even entertaining.) Japan, as many other Asian countries, is very hierarchical in its social and working relationships. This caused a strong cultural shock on my part. It took me a while to get used to the working style and ultimately, the project I was assigned did not bear any fruit during my brief tenure. Nevertheless, when I submitted my report to the funding agency that financed my stay, I asked if I could return, and reapplied for the same fellowship the following year.

During my second visit, I took advantage of the fact that I was already familiar with the laboratory and its facilities. I worked on a project that gave almost immediate results. At the end of my stay, I presented my work in a group meeting, after which the group leader informed me that a paper based on my results would be submitted thereafter. I was quite happy and impressed that I could get a publication from just a couple of month's work. In the months that followed, I tried to keep in touch with the other scientists who were working on the project, and to make myself useful by sending some references that I thought may be interesting for the work. As time passed, I began wondering what had happened to the paper they were supposed to draft after I left. At one point I summoned my courage and wrote to the group leader, asking what the progress was. One of the other collaborators answered me in his place, saying that they had decided to improve the results, and had continued the project where I had left it. He claimed that they had significantly improved the work, but that my contribution to the project at this point was marginal and therefore I only deserved to be acknowledged.

I was baffled. I thought I could trust my collaborators, but evidently they had decided to write me off the project soon after I left. The worst of this story is that there is no way to defend yourself from this type of conduct, except perhaps choose collaborators that you can trust and who will not betray you at the first opportunity.

6.2   R.'s near-fatal M.Sc. experience (Federico)

R. had a terrible experience during his Master's thesis, and when he graduated he was not at all sure that he wanted to continue doing research. R. had crises almost daily, because he felt completely lost and without a sense of direction. After graduating, it took him almost two years and a lot of painful, hard thinking, to decide that he should give himself a second chance, after this first, disastrous experience.

R.'s M.Sc. supervisor, an eminent scientist in his own field, embraces the philosophy that a student should learn to become independent as early as feasible, *basically with as little help as possible from his advisor.* Thus, he would intervene to come to R.'s aid only when it was *absolutely* necessary. Unfortunately, R. was very insecure and his undergraduate training was not nearly good enough to make him independent in the lab when he started his M.Sc. work.

Initially, R. started working on an experimental chamber that was still being commissioned. R. thought it was a very interesting project, but was not experienced enough to realize that it was too far behind schedule for him to benefit from this sophisticated apparatus in the (hopefully brief) course of his M.Sc. experience. After six months of little work and much frustration, R. was eventually moved to another system.

This experimental system had already given some interesting results, which R. found somewhat encouraging. However, R. was now left to work almost totally on his own, and as he began his experiments the chamber almost fell apart on him. It seemed as if it was held together by scotch tape, and R. had to spend most of his days troubleshooting what was broken, and trying to repair it (arguably, R. learned a lot about troubleshooting during this period). More months of frustration ensued, during which R. was finally able to collect enough data to write his M.Sc. dissertation.

At the end, however, R. felt that he had not been properly supervised, that he had wasted a lot of time because of poor organization and bad group management. He was so discouraged that he thought he wanted to give up on his long-term dream of pursuing a scientific career and eventually becoming a professor. Worst still, in spite of all his efforts, there was not a single publication that came out of R.'s thesis. It was particularly frustrating, especially since R. had given much thought about his M.Sc. thesis, had spoken to many different scientists, had considered different options, and ultimately had followed an eminent scientist's advice in choosing his supervisor. In spite of all R.'s thinking, planning and being mentored by an experienced scientist, R. made an initial mistake that seriously jeopardized his career. Despite all the care he had taken in planning ahead and choosing his advisor, R. had an overall terrible experience, bordering on the disastrous.

Alas, as we argued before, luck plays a very important role in human matters. Unfortunately, even planning ahead may not always protect you from fortune's misdeeds.

6.3   T.'s case: Insecurity and stubbornness can be fatal (Federico)

T. was one of the best students of my university course (he had an average of 29.8 out of 30, almost a record in the Italian university system), and a good friend of mine. We used to sit next to each other throughout all the classes in the third and fourth years of University.[a] Just like all the other bright students in our course, T. decided to become a theoretician.[b] He seemed to be particularly fascinated by the material we learned in the courses on Statistical Mechanics and Superconductivity, and decided to do his M.Sc. thesis with one of those two professors, who were actually collaborating on most of their research projects. He certainly had the ability to learn the subject material very rapidly, and to present it superbly. He told me much later that he had made up his mind *following the choice of others, because he thought he was at least as good as the best students in the course.* I hope it is clear enough how silly this choice was, without having to explain it in more detail. I did try to warn him, and strongly suggested to him to look into other

---

[a]I recall a particular lesson on Quantum Mechanics, in which the whole class was completely lost. Even T. seemed to be lost, since he was muttering and complaining while taking notes. I felt somewhat relieved, because I thought that if even he did not understand, I might as well archive the content of the lesson, and move on. At one point however, he raised his hand and asked the instructor: "Shouldn't there be a minus sign in that equation?" We were about to kill him. Not only he had understood everything, he was even able to spot a mistake in the calculations.

[b]I would like to digress a bit here. The first day I saw T., he sat next to me in the course on "Mathematical Methods for Physics." He had long hair and an unshaven face, and he spent the whole interval between lectures tidying his notes with his incomprehensible handwriting. He made such a bad impression on me that I thought he must not be particularly smart. He turned out in fact to be one of the very best students I ever met, and in some sense I still believe he is a genius. This showed clearly to me that appearances can easily deceive. A scientist should not make snap judgments from superficial appearances. I should have known better.

possibilities. He went to talk to other professors only half-heartedly, having already decided that he wanted Dr. SuperFamous and friend to supervise him *because they were the best* (at least in his opinion) and because the best students had migrated towards them.

The result was an unbelievable disaster. Dr. SuperFamous and friend held T. in high esteem, because they knew he was a very bright student. They had both been particularly impressed by his performance in their exam. So they did not worry too much when they did not see him come to knock on their door every other day — which is what the other M.Sc. students were doing. They thought he could work it all out by himself. On the other hand T. was too shy to go disturb his advisors on a regular basis. He was also very stubborn, and had decided early on that he should be able to solve most of the problems he faced independently, without any help from his advisors. Although he was aware of being a very bright student, paradoxically he was very insecure. By the end of his thesis, T. had interacted little with his supervisors, had learned very little from them. What is worse, he had accumulated an enormous amount of frustration and insecurity.

Later T. got a fellowship for his Ph.D. work, and again made the mistake of going back to the same advisors (in spite of repeated attempts on my part to dissuade him). It did not take a great brain to see where this was going. Less than a year into his Ph.D., he received an offer to become a consultant, and his scientific history ended before it even began. I suppose T. was too stubborn and perhaps too insecure to pursue a scientific career. Probably he even lacked the drive and the passion which are the essential motors that keep most scientists going. However I am convinced that if he had chosen his advisors more wisely, he would have at least avoided a great deal of frustration, and perhaps would have had a shot at becoming a scientist.

Today T. claims that he is happy with his employment, which is more or less a nine-to-five job that leaves him plenty of time to spend with his wife and do other things. Perhaps he did the right thing in the end. Sometimes, however, I still wonder if we lost a great scientific mind (much better than mine), essentially because of poor planning and naïve choices.

6.4   M.'s case: Half-hearted decisions are unwise (Federico)

M. had a great passion for astrophysics. She was a very bright student, and pursued her M.Sc. thesis with great enthusiasm and energy. It took her longer than normal to graduate, but she did not intend to let this affect her future career plans. Although her interaction with her advisor was not always positive, her spirits were high about the kind of work she was doing. She was driven by true passion, the driving force of all scientists. When she was about to graduate, she received an offer to continue her graduate studies at the University of Toronto. It was a difficult choice at best, since her boyfriend was living and working in Rome, and was not likely to move to Canada with her, at least not at the beginning. At the same time, she was very attached to her family and did not want to leave her personal life behind. After a lot of hard thinking, she decided to give it a try. When she was about to leave, I predicted[c] that she would come back before finishing her thesis, and that it was just a matter of time. Indeed, she gave up only about ten months into her Ph.D. Her decision to leave behind her family, boyfriend and personal life was half-hearted. Perhaps she secretly hoped that her boyfriend would look for a job and eventually move to Toronto. When it became apparent that this would not happen, she decided to give up. Now she works for a mobile telephone company in Rome, and her dream of doing research in astrophysics has vaporized, probably forever.

6.5   R.'s Ph. D. experience (Federico)

After R.'s disastrous experience as an M.Sc. student (Sec. 6.2), it was not easy to find the courage to start over and tackle a Ph.D. project. When R. received a Ph.D. fellowship from the University of Rome "La Sapienza," he was faced with the following options: either go back to his M.Sc. supervisor's laboratory, hoping that things would work out this

---

[c]For some reason that is still unclear to me, I can often predict when someone I know well is about to make a huge mistake in a personal or career decision. The sad reality however, is that people rarely listen to me when I warn them about the blunder they are about to plunge into. I hope the readers of these notes will be wiser and will be able to avoid some mistakes and a lot of frustration.

time (after all he had eventually learned to use the techniques and he knew exactly what to expect, or rather, what not to expect), or look for a completely new project, which would obviously be more fascinating and challenging, but also more risky.

R. started looking around and visited several labs in the three Roman Universities. In his search, he used various parameters to optimize his choices. Among these, he looked for an advisor who was likely to invest time in his supervision rather than leaving him completely on his own; and for a research topic that would give him a good chance of landing a job once his studies would be completed. This is another important piece of advice. Do not follow the crowd, but rather, try to pick projects and pursue ideas that will turn you into a leader. If you succeed in doing this, jobs will chase you, instead of the other way around.

He talked to several people who had interesting projects to offer, and did a lot of background reading to get a better idea of their field of research and where this specific project would come in. In the end however, R. had one particular conversation with a senior professor that he found particularly convincing and inspiring.

When R. walked out of his door after a one-on-one interview he thought, *"today I had the rare privilege of talking to a real SCIENTIST."* This person, who later became R.'s advisor, described with enthusiasm and with great clarity his interests and the type of work he would be doing if R. joined his group. His description was so seductive that even if R. told him that he would think about it (which he did), R. already knew that he would take his offer. *A posteriori*, R. could not have chosen any better than that in a million years!

Finally, in this new laboratory R. was properly supervised. He learned new techniques, including how to use the Scanning Tunneling Microscope, a fancy experimental technique with which R. soon started taking images of surfaces with atomic resolution.[d] He tackled this new project with great enthusiasm, and has been riding a wave of success ever since.

---

[d] Since then, the description of R.'s job to the layman is, "I play with atoms and molecules, and actually get paid for it."

Remember, when you finish your Ph.D., you will be called a Doctor (from the Latin verb *doceo*, which means to teach). You will have reached the highest level of education offered anywhere in the world (except for those crazy people who have more than one doctorate). It is a time to celebrate, because you have achieved something that few people can achieve. In many ways, you should be proud of yourself. After you have celebrated adequately, you should remember that this is a milestone to build upon, and not an end in itself.

6.6   R.'s post-doctoral experience (Federico)

When R. was hunting for a post-doctoral job, towards the end of his Ph.D. thesis, he received several offers over a period of about 10 months. In fact R. applied for a second round of jobs when he had already started his post-doctoral work, because he was having problems in adjusting to the new group.

During the first round, R. received one offer from Sweden (which is the job he ended up accepting), two offers from Korea, two from Italy, and one from the United States. When sending out his applications, for whatever reason, R. was old-fashioned and decided to send them all by ordinary mail rather than by email. (The net effect of this silly choice was that some applications got lost in the mail, and many arrived several months after the deadline had expired.)

Knowing the Italian system fairly well, in all its negative aspects, R. never took the Italian offers very seriously. So what motivated him to go to Sweden instead of the U.S. or — God forbid — the remote and mysterious Korea? The offers he got from the United States and Korea all came by email.

In fact, R. was offered the two positions in Korea during an exchange program in Japan. Since he was not too far from Korea and essentially on the same time zone, R. expected at least a phone call from the labs that claimed they wanted to employ him. (R. was actually hoping that they would invite him over for an interview, but did not have the courage to ask.) It would have been costly, of course, but since he was nearby he thought it made perfect sense to take advantage of the situation to meet personally, also giving him the opportunity to visit the country (I completely agree with him). None of this happened.

To R.'s dismay, his prospective boss in the United States (Texas, to be precise) behaved essentially in the same way. Reading closely the messages he was sending, it became clear that he was not interested in hiring R. as a person, but rather as a form of *cheap labor*. He wrote upfront that in his group people worked on average 6 days a week, for a total of 60 something hours, and that they would take seven to ten days of paid vacation per year. (Incidentally, this is a typical work schedule in many different environments in the United States, where the system tends to exploit people whenever and as much as possible.) This is not surprising and is not even uncommon, and it is exactly what R. has been doing all along, during his Ph.D., his post-doctoral work in Sweden, and even more so now that he is a young professor. However, R. found it distasteful to be told all this upfront. Working that hard is a matter of personal choice, something that one chooses to do to advance his/her own career. It is certainly not something that you should do simply to please your employer.

In contrast to all this, the Swedish professor invited R. to an interview. He generously offered to reimburse all the travel costs (which I personally think should be the norm, but unfortunately a lot of stingy colleagues would disagree). R. had the opportunity to spend about 3 days in the professor's town in Sweden, visiting his laboratory, talking to other members of the group, and getting a general impression of what it would be like to live there for a couple of years. It was the end of May and in Rome R. was wearing a T-shirt, but in Sweden it was still very cold and he had to wear a sweater. R. recollects vividly that this weather factor scared him quite a bit, and he kept telling myself that he would never go to work there (I can relate to that — I live in Montreal, where the temperature goes down to minus thirty Celsius during the winter!).

During a one-on-one interview with the Swedish professor, R. was told explicitly that the professor always wanted to meet the people he was going to hire, if it was at all possible (he probably did not interview prospective candidates from China or Japan, unless he would meet them during a trip to the east). R. actually thought it made sense. (I completely agree with this philosophy, and, whenever feasible, I do the same.) R.

also realized that, if you are good enough to receive more than one job offer, you will be able to choose, and in that case attending an interview is a situation in which both parties are trying to please each other.

After a few months in Sweden however, R. was seriously thinking of leaving, particularly because he did not get along very well with the graduate student he was supposed to work with. In hindsight, their personalities and views were quite different, so it was an improbable match at best (although of course they had not realized this when they first started working together). Even worse, the student was one year older than R., and still had to finish his Ph.D., whereas R. was employed there as a post-doc. This, among other issues and misunderstandings, created a lot of friction, and eventually prompted R. to send out a second round of applications.

When he received replies from this second round, he made it a point to take seriously only those groups that were inviting him over for an interview, and that were offering to pay for his travel expenses. R. ended up receiving two more offers, one from Halle (Germany) and one from Zürich (Switzerland). In the end he was not impressed by either of the two places, not scientifically, but because he did not like the town where he would have to live. Ultimately also the place where you live is important, because although you are going to work long hours in the lab, you will undoubtedly need something interesting to do outside of working hours.

Before deciding however — I forget, but perhaps R. was close to choosing the offer from Switzerland — R. had an open conversation with his post-doctoral advisor, the Swedish professor I mentioned before. R. explained his situation, and the boss convinced him to stay. R. felt kind of silly at that point, because he realized that probably most of the problems he was facing were due to a lack of openness and communication between himself, the graduate student and the boss. The situation improved steadily and substantially from then onwards, and in retrospect R.'s staying in Sweden was a very wise choice, that helped his career immensely.

Incidentally, now that I am the "boss," whenever I receive an application from a prospective and promising student, I make it a point to invite him/her over for an interview whenever it is feasible.

(Unfortunately I do not have the means to invite candidates from overseas, and even if I did, I am not sure it would make sense to go through all this trouble at the level of a Ph.D. student.) If a live interview is not possible, I usually schedule a telephone conversation instead. This is exactly what I did with my current students — either met them in person or interviewed them on the phone.

6.7   An unconventional career: Thinking outside the box

During a conference in a developing country I met L., a European scientist who is presently working in Asia. L. has a very unconventional approach to scientific research, and we are therefore convinced that it is interesting to sketch his career and his views for the benefit and perhaps the inspiration of a broader audience. From this excerpt of Federico's interview to him it will become apparent that his approach is quite uncommon, that it is not suitable for the majority of people. However, some young scientists who are ambitious and have a clear vision may want to follow in his footsteps.

L. is considered a "senior" scientist in his field, in the sense that although he is still fairly young he has made important contributions and advances (and he has earned his doctorate a while ago). At this stage in his career, he could easily find a permanent position — if he wanted to. The surprising thing is, he does not want to at all. He claims — with reason — that once you acquire a permanent job, you are tied down to the rules and dynamics (very slow, more often than not) of your institution. This tends to slow you down considerably and to hamper your overall progress in research.

Although we presume he would not like to be categorized, in our description of scientific research L. really is a beta, but an unusual beta indeed. He likes to be the one who turns the knobs. He is not particularly interested in training students; rather he sees them as an impediment to his progress. He likes doing all the research himself, from A to Z, and does not want to deal with anything that is going to waste his time and hamper his progress.

L. is a scientist who likes to think "outside the box," and so far has based his career on quite unconventional strategies and approaches. He seeks work environments and job opportunities where there are essentially no limits to the freedom granted to conduct research.

L. does not want to be "encapsulated" or worse, tied down into a "permanent" position. According to him, the perceived advantage of having some form of job security comes at a price which is too high because it imposes the long set of rules of the institution that retains you. In many cases he believes that a permanent position is psychologically bad, because it can easily drive people into boredom — i.e. job security for some people can be de-motivating. (If we take this view on a case-by-case basis, we tend to agree fully. In fact the converse may also be true: permanence can be seen as a form of professional recognition, and not granting it may de-motivate people who think they deserve it. So much depends on the individual.)

Thus L. has been hopping between various positions as "visiting scientist," or visiting professor. Sometimes he takes on an offer for as little as six months, then moves straight to the next opportunity. These positions, more often than not, entail "special" treatment on the part of the host laboratory, including e.g. easy and fast access to resources and facilities — which may be a lot more difficult for the scientists "employed" formally at the same institution.

Since it is hard for anyone to tackle an ambitious project in a short period of time, the P.I.s who invite L. give him maximum freedom and almost immediate access to anything he may need or ask for. Thus, L. maintains that for visiting scientists the "general rules" do not apply. Precisely because the overall visiting period is limited, the urgency allows L. to negotiate more efficient working conditions.

In this way, so far L. has been able to defy conventions and traditional approaches, and to constantly challenge himself — which is what doing science is all about. He is effectively forging his own career and way of life at the same time.

L. is constantly pushing himself to be more imaginative, creative and of course, organized. Whenever he takes up a new offer, he comes in with his own project, a clear idea of what he wants to do, and makes sure that everything is in place before he arrives. His efficiency is maximized also because the overhead (in terms of time) of securing the funds rests on someone else's shoulders, i.e. the P.I. or the manager of the lab who invites him.

L. acknowledges that this strategy cannot be adopted by anyone. You have to know exactly what you want, be ready to move frequently, work very efficiently, etc. In essence, not everybody is up to the challenge, and of course if you have a family it becomes next to impossible. But if you are of a very special character, this unconventional (and very uncommon) modus operandi can be very stimulating and rewarding.

Among his various activities, L. is also Editor of a new journal in his field of research. What is remarkable is that he negotiated this position in such a way that he would not be paid for his services — again, if he were formally "hired" and paid, this would limit his freedom to implement his personal vision for the journal.

L. prefers working in national (government) laboratories, because the focus is on doing research, not teaching. He does enjoy giving seminars and lectures, but only to an interested audience. In a University, L. assumes (perhaps correctly) that during formal lectures only a small minority of the audience is actually interested in listening and learning. In essence, L. does not want to make any compromise with the use of his time; for example, he does not like to waste time doing administrative work (well, after all who does...?) and he does not like the very concept of supervision, as he considers it a waste of time. He never went to work "for" someone, because he always brings his own project with him. He does not want to train or supervise students, because he does not want to make "copies" of himself.

His unique approach and his high profile in research so far allowed L. to "create positions" for himself (at least the types of positions he is interested in).

L. also has some very specific views about ethics. For example, he does not allow spectators to use cameras or video cameras during his talks — essentially for fear of plagiarism and improper use of his material.

Chapter 7

L'Envoi

L'envoi — the definition: One or more detached verses at the end of a literary composition, serving to convey the moral, originally employed in old French poetry.

Pursuing a scientific career is a very challenging endeavor, and, as such, it is not meant for everyone. It requires a great determination to succeed, a lot of patience, and much dedication and perhaps even sacrifice. It presupposes, in our opinion, that you are extremely enthusiastic about your work, and actually take great pleasure in most of its aspects. This enthusiasm will help you get through the rough times that you are likely to face, at least once in a while. In many ways, you should consider this a mission, or perhaps even a higher calling. When you get up in the morning, you should be anxious to start a new day of exciting, fun/work. It is a rare privilege to be paid to do something you really love. If you do not feel the same way about being a scientist ... well, as we have suggested before, you should seriously consider alternative career plans.

While this is all very well, you should also open your eyes to the real world in which this game of science is played out. Being naïve in a world filled with sharks can be extremely dangerous. (Incidentally, it is very important to realize that each job — indeed, each human enterprise — has a political aspect to it, which must be dealt with skillfully, carefully, and diplomatically.) If you do not understand the game you are in, you are not giving your talent its full scope. In our view, because many young scientists begin with very starry-eyed visions of the world of science, this is particularly true in a scientific career, whether you choose to pursue it in industry, in academia, or in a government laboratory.

We repeat here our main piece of advice. If you want to succeed in this career, plan ahead, carefully and thoughtfully. Most of your plans will be derailed by the events, but if you have done your homework and have thought about your position on the chessboard of science, your chances of being successful or even just surviving as a scientist will be much higher than if you merely let yourself be overtaken by events.

This is our "zeroth law" of scientific survival: pay attention to your scientific career.

As we see it, there are (in the natural order of application) three laws for survival in science.

The first law we give is "Know thyself." Since you are acting as your own agent, as such, you (the agent) must have a clear idea of what you (the client) are and, given that what activity in science is to be your goal. The main thrust of this first law is explored in Chapter 1.

The second law is "Know your tradecraft." Most of this book is devoted to this subject. Chapter 2 discusses many of the basic actions you can take as a "player" in this "game of science" and the overall setting, while Chapter 3 focuses on the specific rules of the "science game". Chapter 4 discusses in detail how build your reputation in the optimum way, given your particular level of scientific ability, while Chapter 5 discusses the all-important craft of writing as a basic tool.

The third law is "Know thy neighbor." When the game of science is being played in a scenario of anonymous peer review (as it often is), this third law does not apply. But when the other players are known your tactics need to be tailored to these people. This theme is scattered through the anecdotes (Chapter 6) and in the discussions and examples of the previous chapters.

We firmly believe that this book will help you focus your ideas so that your chances of success will be considerably higher than if you do not work on applying these principles.

Good luck and have fun in science! We wish you all the best. We invite you to send us your reactions to this work.

Bibliography and Further Reading

P.J. Feibelman, *A Ph.D. Is Not Enough: A Guide to Survival in Science*, Addison-Wesley, Reading MA, USA (1993).

M. Anderson, "*So, you want to be a professor!*", *Physics Today* (Special Issue: "Careers for Physicists") **54** (4) 50-54 (April 2001).

Besides the bibliography in the references, other useful snippets can be found in *Nature* and in *Naturejobs*, similar to those indicated below.

Naturejobs

"*Private foundations push for higher post-doc salaries*", Naturejobs, 10 January 2002.

"*At last, a chance for postdocs to learn how to teach*", Naturejobs, 28 February 2002.

"*Easing the journey back home*", Naturejobs, 28 March 2002.

"*Getting into good company*", Naturejobs, 25 April 2002.

"*Training programs with vision*", Naturejobs, 23 May 2002.

"*Equal opportunities*", Naturejobs, 20 June 2002.

"*Working your way into industry*", Naturejobs, 5 September 2002.

"*Does travel broaden the scientific mind?*" Naturejobs, 17 October 2002.

"*Seeking strength in numbers*", Naturejobs, 9 January 2003.

Nature

"Getting organized", Nature **423**, 98 (2003).

"Location", Nature **421**, 766 (2003).

"Postdoc positions axed", Nature **420**, 452 (2002).

"Stacking the deck", Nature **422**, 784 (2003).

Interesting links to some websites:

Advice to the Young Astronomer by Ed Nather:
http://whitedwarf.org/education/advice/index.html

You and Your Research by Richard Hamming:
http://whitedwarf.org/education/hamming/index.html

Ethics in Science website:
http://www.towson.edu/users/sweeting/ethics
/ethicbib.htm#B.ScientistsExperiments

Glossary

Article:
This is the "classical" type of scientific publication. It normally consists of a title, author list and abstract, followed by a main text. The main text in turn consists of an introduction, an experimental/methods section, a section on results, possibly a discussion section, and finally a concluding paragraph. The latter is followed by acknowledgements and a list of references. Figures with their legends/captions are usually embedded in the main text.

Brain Drain:
In a dictionary definition, brain drain means the *loss* of highly qualified people by emigration.

Brain Gain:
In a dictionary definition, brain gain means the *gain* of highly qualified people by immigration.

Chair, Endowed Chair:
In a dictionary definition, a chair is a seat of authority or office, a professorship.

Citation Index:
After the invention of the web it became significantly easier to monitor scientists' citations. There are now commercial indexes that can measure the total number of citations of a scientist as well as the impact factor of a journal (which is an essence an average number of citations per paper published in the journal over a given period of time). Citation indices currently in use include those from the ISI series (ISI Web of Science and ISI Web of Knowledge).

Communication:
A scientific "communication" is very similar to a Letter. Depending on the journal, communications may or may not have an abstract. These forms of "short" publications are becoming increasingly competitive. In a world where scientists are becoming increasingly busy, researchers claim that they do not have the time to read long articles and therefore focus only on reading letters and communications. As a result, anybody who wants to be widely read (and cited) will try to publish in these specific sections of a journal.

Curriculum Vitae (CV):
Also called "résumé" (especially in business circles), the curriculum vitae is a Latin expression which means "the course of studies in your life."

Fellowship:
See *scholarship*.

Grant:
In a typical dictionary a grant would be defined as "a sum of money bestowed or allowed". In scientific research, a grant is a sum of money awarded by a funding or granting agency or by a foundation to individual or groups of scientists to conduct specific research projects.

Impact:
In everyday life, "impact" usually refers to something unpleasant, like for example a collision between objects and/or people. In scientific research, saying that a scientist has had an "impact" means that this researcher has left a significant mark in his/her field of study. Evidence of impact comes in the form of citations to a scientist's work, invitations to present his/her work, awarding of prizes, fellowships, grants, honorary degrees, commercialization through patents, etc.

Letter:
In the context of scientific publications, Letters (sometimes "Letters to the Editor") are typically short articles. They are usually processed faster

by the journal, both at the peer review and at the production levels, because they are considered more "urgent." In principle, they are expected to have a higher impact with respect to "regular" or "full" papers.

Mentor:
In a dictionary definition, a mentor is a faithful guide, or a wise counselor.

Patent:
In a dictionary definition, a patent is defined as the exclusive right to make or sell a new invention.

Scholarship:
In a dictionary definition, a scholarship refers to "emoluments so granted to a scholar," or to "education, granted to a successful candidate after a competitive examination."

Acronyms and Funding Agencies

The list that follows is designed to help young students and scientists find information about fellowship, funding and prize opportunities. It is neither complete nor exhaustive.

<u>United States of America</u>

National Science Foundation (NSF):
www.nsf.gov

National Institute of Health (NIH):
www.nih.gov

Department of Energy (DOE):
www.energy.gov

Department of Defense (DOD):
www.defenselink.mil

Army Research Office:
www.aro.army.mil

Air Force Office of Scientific Research (AFOSR):
www.afosr.af.mil

Office of Naval Research:
www.onr.navy.mil

DARPA:
www.darpa.mil

Canada — Federal Government

Natural Science and Engineering Research Council (NSERC):
www.nserc.ca

Canadian Institutes for Health Research (CIHR):
www.cihr.ca

Canada Foundation for Innovation (CFI):
www.innovation.ca/index.cfm

Canada Research Chairs (CRC):
www.chairs.gc.ca/web/home_e.asp

Canadian Space Agency:
www.space.gc.ca/asc/eng/default.asp

CSA Materials Science Program:
www.space.gc.ca/asc/eng/csa_sectors/space_science/microgravity/
material.asp

Department of Foreign Affairs and International Trade (DFAIT):
www.dfait-maeci.gc.ca/menu-en.asp

Canadian International Scholarship Programs:
www.scholarships-bourses-ca.org/menu-en.html

Commonwealth Scholarships:
www.scholarships-bourses-ca.org/pages/CWin/aCW_ToCan1-en.html

Scholarships from the Organization of American States (OAS):
www.scholarships-bourses-ca.org/pages/OASin/aOAS_Ca_in1-en.html

Government of Canada Awards:
www.scholarships-bourses-ca.org/pages/GCAin/aGCA_in1-en.html

Killam Research Fellowships:
www.canadacouncil.ca/prizes/killam/xy127235773746406250.htm

The Royal Society of Canada:
www.rsc.ca

Canada — Provincial Governments

Quebec

Fonds Québécois pour la Recherche en Nature et Technologies (FQRNT):
www.fqrnt.gouv.qc.ca

Fonds pour la Recherche en Santé du Québec (FRSQ):
www.frsq.gouv.qc.ca/fr/index.shtml

Ministère pour le Développement Économique Régional et pour la Recherche (MDERR):
www.mderr.gouv.qc.ca

Valorisation Recherche Québec (VRQ):
www.vrq.qc.ca

Ontario

Materials and Manufacturing Ontario (MMO):
www.mmo.on.ca

Photonics Research Ontario (PRO):
www.pro.on.ca

Ontario Research and Development Challenge Fund (ORDCF):
www.ontariochallengefund.com

Europe: European Union, European States

European Science Foundation (ESF):
www.esf.org

Science Foundation Ireland (SFI):
www.sfi.ie

Marie Curie Actions:
http://europa.eu.int/comm/research/fp6/mariecurie-actions/action/
level_en.html

Asia/Pacific/South Pacific

Japan
Japanese Society for the Promotion of Science (JSPS)
www.jsps.org

Australia
International Scholarships and Exchanges Funded by the Australian
Government:
www.dest.gov.au/International/Awards/endeavour.htm

Australian Research Council:
www.arc.gov.au

Singapore
A*: Agency for Science, Technology and Research (A-STAR):
www.a-star.edu.sg/astar/home.do

Asian Office of Aerospace Research and Development (AOARD, part of AFOSR):
www.tokyo.afosr.af.mil

National/International Foundations and Companies

Alfred P. Sloan Foundation:
www.sloan.org/programs/index.shtml

Guggenheim Foundation:
www.gf.org/broch.html#elig

The Human Frontiers Science Program:
www.hfsp.org

Fondazione della Riccia (Italy):
http://arturo.fi.infn.it/casalbuoni/dellariccia

Research Corporation:
www.rescorp.org

The Volkswagenstiftung (Volkswagen Foundation):
www.volkswagen-stiftung.de/english.html

The Canon Foundation:
www.canonfoundation.org

Camille Dreyfus Teacher-Scholar Award Program:
www.dreyfus.org/tc.shtml#eligibility

The MacArthur Foundation:
www.macfound.org

The Semiconductor Research Corporation (SRC):
www.src.org/member /about.asp?bhcp=1

SEMATECH:
www.sematech.org

The Beckman Young Investigator Award:
www.beckman-foundation.com/byiguide.html

The Dupont Young Faculty Award:
www.dupont.com

David and Lucile Packard Foundation:
www.packard.org

The Don & Sybil Harrington Foundation (Texas):
www.rra.dst.tx.us/c_t/people/DON%20AND%20SYBIL%20
HARRINGTON%20FOUNDATION.cfm

The Welch Foundation (Texas):
www.welch1.org

Hellman Family Faculty Award (Berkeley, California):
http://vpaafw.chance.berkeley.edu/hellman.html

Prizes and Awards

The Steacie Prize:
www.steacieprize.ca/index_e.html

The Royal Society of Canada: Medals and Awards:
www.rsc.ca/index.php?&page_id=61&lang_id=1

Materials Research Society (MRS) Awards:
http://www.mrs.org/awards
(Von Hippel Award, David Turnbull Lectureship, MRS Medal,
Outstanding Young Investigator Award, Graduate Student Awards)

European Physical Society (EPS) Prizes:
www.eps.org/prizes.html

American Physical Society (APS) Prizes and Awards:
www.aps.org/praw/index.cfm

McMillan Award for Condensed Matter Physics:
http://web.physics .uiuc.edu/General_Info/McMillan

Professional Societies

Professional Societies play various important roles in the Scientific community. Among other tasks, they organize annual meetings and conferences in disciplines relevant to the Society, distribute prestigious prizes and awards, and publish journals in fields of interest to the Society.

Materials/Physics/Chemistry Professional Societies in the United States

Materials Research Society (MRS):
www.mrs.org

American Physical Society (APS):
www.aps.org

American Institute of Physics (AIP):
www.aip.org

American Chemical Society (ACS):
www.chemistry.org

American Vacuum Society (AVS):
www.avs.org

Materials/Physics/Chemistry Professional Societies in Europe

European Physical Society (EPS):
www.eps.org

European Materials Research Society (E-MRS):
http://www-emrs.c-strasbourg.fr

Materials/Physics/Chemistry Professional Societies in Canada

Canadian Association for Physicists:
www.cap.ca

Canadian Society for Chemistry:
www.chemistry.ca

Professional Journals and Magazines

Physics Today:
www.aip.org/pt

Chemical and Engineering News:
http://pubs.acs.org/cen/index.html

Materials Research Bulletin:
www.mrs.org/publications/bulletin

Physics World:
www.physicsweb.org

Nanotechnology Portal:
www.nanotechweb.org

Nature Jobs:
www.nature.com/naturejobs

Science's Next Wave:
http://nextwave.sciencemag.org

References

[1]S. Harris, *Chalk Up Another One* (Rutgers University Press, 1992). Of his many books, this is the one we cited the most. Others may be found on his Web site (www.sciencecartoonsplus.com) and at Amazon.com.

R.L. Weber's two anthologies *A Random Walk in Science* (Institute of Physics, 1973) and *More Random Walks in Science* (Institute of Physics, 1982) (the one most cited here).

S.G. Krantz, *Mathematical Apocrypha* (The Mathematical Association of America, 2002). Great fun to skim, although many of the best anecdotes are to be found originally in *Littlewood's Miscellany*, ed. Belá Bollobas (Cambridge University Press, 1986).

Science Jokes also abound at the Web site www.buzzle.com/chapters/science-and-technology_jokes-and-funnies.asp, e.g. *The Game of Refereeing* can be downloaded there.

[2]P.J. Feibelman, *A Ph.D. Is Not Enough: A Guide to Survival in Science* (©1993 Peter J. Feibelman) (A Web-search (e.g. Google) will find it on Amazon.com, or you can go there directly.)

[3]R.M. Reis, *Tomorrow's Professor: Preparing for Careers in Science and Engineering* (Institute for Electrical and Elecronics Enginieers, 1997).

D.F. Bloom, J.D. Karp and N. Cohen, *The Ph.D. Process: A Student's Guide to Graduate School in the Sciences* (Oxford University Press, 1998). (Good for undergraduates going to graduate school.)

K. Barker, *At the Helm: A Laboratory Navigator* (Cold Spring Harbor Laboratory Press, 2002). (Excellent if you are faced with setting up your own experimental laboratory.)

[4]A.J. Friedland and C.L. Folt, *Writing Successful Science Proposals* (paperback) (Yale, 2000).

T.A. Ogden and I.A. Goldberg, *Research Proposals: A Guide*, 3rd ed. (Academic Press, 2002).

[5]M. Alley, *The Craft of Scientific Presentations: Critical Steps to Succeed and Critical Errors to Avoid* (Springer-Verlag, New York, 2003).

R.A. Day, *How to Write and Publish a Scientific Paper*, 4th ed. (Oryx Press, Phoenix, 1994).

J.R. Matthews, J.M. Bowen and R.W. Matthews, *Successful Science Writing: A Step-by-Step Guide for the Biological and Medical Sciences* (Cambridge University Press, Cambridge, 2000).

S.L. Montgomery, *The Chicago Guide to Communicating Science* (University of Chicago Press, Chicago, 2003).

J.A. Pechenik, *A Short Guide to Writing about Biology*, 5th ed. (Harper-Collins College Publishers, New York, 2004).

G.M. Whitesides, Whitesides' Group: Writing a Paper, *Advanced Materials* **16**, 1375 (2004).

[6]S. Goodlad, *Speaking Technically* (Imperial College Press, 1996).

J.C. Garland, Advice to Beginning Physics Speakers, *Physics Today*, July, 42–45 (1991).

[7]M.H. Briscoe, *Preparing Scientific Illustrations: A Guide to Better Posters, Presentations, and Publications*, 2nd ed. (Springer-Verlag, New York, 1996).

S. Block, The DOs and DON'Ts of poster presentation, *Biophysical Journal* **71**, 3527–3529 (1996).

D.A. Keegan and S.L. Bannister, Effect of colour coordination of attire with poster presentation on poster popularity, *Canadian Medical Association Journal* **169**, 1291–1292 (2003).

C. Rigden, 'The eye of the beholder' — Designing for colour-blind users, *British Telecommunications Engineering* **17**, 2–6 (1999).

E.R. Tufte, *The Visual Display of Quantitative Information* (Graphics Press, Connecticut, 1983).

T.G. Wolcott, Mortal sins in poster presentations or, How to give the poster no one remembers, *Newsletter of the Society for Integrative and Comparative Biology*, Fall, 10–11 (1997).

J.D. Woolsey, Combating poster fatigue: How to use visual grammar and analysis to effect better visual communications, *Trends in Neurosciences* **12**, 325–332 (1989).

Index